Louis D. Ricketts

The Ores of Leadville and Their Modes of Occurrence

As illustrated in the Morning and Evening star mines, with a chapter on the methods of their extraction as practiced at those mines

Louis D. Ricketts

The Ores of Leadville and Their Modes of Occurrence
As illustrated in the Morning and Evening star mines, with a chapter on the methods of their extraction as practiced at those mines

ISBN/EAN: 9783337036232

Printed in Europe, USA, Canada, Australia, Japan

Cover: Foto ©berggeist007 / pixelio.de

More available books at **www.hansebooks.com**

THE ORES OF LEADVILLE

AND

Their Modes of Occurrence

AS ILLUSTRATED IN THE

MORNING AND EVENING STAR MINES,

WITH A CHAPTER ON THE

Methods of their Extraction as Practiced at those Mines,

BY

LOUIS D. RICKETTS, B.S.
Ward Fellow in Economic Geology of the College of New Jersey.

WITH FIVE PLATES AND ONE COLORED LITHOGRAPH.

PRINCETON, 1883.

PREFACE.

In accordance with the requirements of the W. S. Ward fellowship in Economic Geology for the year 1882-3, I spent over four months of that year at Leadville, and devoted my time to a study of the ores and their modes of occurrence, and to the extraction of ores in the Evening and Morning Star mines.

To these mines I had free access, and would here tender my thanks to their officers, who afforded me every facility in their power for the prosecution of my work. The analyses of the rocks and ores were made and the thesis completed at the School of Science, Princeton, during the remainder of the college year.

I have been greatly stimulated throughout by the personal interest taken by Mr. Ward in the work; and it is at his request and expense that this thesis, written in compliance with the terms of the fellowship, is published.

Part First is also presented as a thesis for the degree of Doctor of Science.

L. D. RICKETTS.

PRINCETON, N. J.
June, 1883.

PART FIRST.

THESIS FOR THE DEGREE OF DOCTOR OF SCIENCE.

INTRODUCTORY.

This thesis does not so much attempt a general description of all the mines of Leadville as a more particular description of the mines to which the writer has had free access for several months. And it is hoped by a more detailed description of these quite extensive mines to give a better idea of this type of deposit than could be done by a more superficial description of a greater number of mines. Other mines, both on Carbonate and Iron Hill, have been visited in order that common features may be noted and peculiarities distinguished, but almost all the data for the following pages have been derived from the MORNING and EVENING STAR MINES.

The Morning Star especially has been studied, for this mine is only at present undergoing development, and large bodies of ore stand open, completely explored by drifts but otherwise untouched, thus affording most excellent opportunities of examination.

The Evening Star, on the other hand, is much smaller, and the ground developed is practically exhausted. Its enormous ore-body has been removed, and only lofty cribbing and timbers, reaching up set over set, show the great spaces once occupied by almost pure ore. The present shipments (October and November, 1882) are obtained from the smaller ramifications of the ore through the gangue, and the apparently interminable number of these smaller bodies promises to be the source of much more ore. Further prospecting work is now being pushed forward, through which it is by no means impossible that new ore-bodies may be struck. It would be truly rash to assert that the Evening Star mine is nearly exhausted without adding the qualifying clause, if new ore-bodies are not presently developed.

The excellent report of Emmons, an abstract from his final report on the geology and mining industry of Leadville, has been of the greatest assistance to the writer, and is used as an authority in the short description of the general geology deemed necessary for an intelligent comprehension of the subject treated. For a fuller description of the general geology that pamphlet is referred to, as space for this is lacking here. The statements of Emmons concerning the strata and their nature at different horizons have corresponded very well to the facts gathered in the limited field here treated. As will be mentioned later, the position of the Carbonate fault is found to be further down the hill than it is stated to be by that authority, and indeed no direct evidence of its existence could be obtained on the property. These and a few other points of minor importance rendered apparent by the developments lately made on the lower part of the properties disagree with his statements. All the others correspond very nearly.

GENERAL GEOLOGY OF THE DISTRICT.

The stratified rocks about Leadville are almost altogether Palæozoic, with an immense amount of Quaternary débris on the lower slopes of the hills, and running into and over the broad valley of the Arkansas. Between certain of these stratified Palæozoic beds are found a complicated series of interlaminated intrusive sheets of the igneous rocks so intimately connected with the ore-deposits. The upper part of the Mosquito Range and of the opposite Sawatch Range is composed of Archæan rocks, principally granites and gneisses. These are of dull colors, for the most part gray and pinkish. Gneisses predominate, and in these a distinct bedding is often recognizable. They split up and disintegrate along these lines as well as along the jointing planes. Frequently, along the Sawatch Range at least, they are intersected by dikes of igneous rock which weather to thin plates and blacken and form prominent stripes of slide down the mountain-side wherever they occur.

The sedimentary beds form a belt around the base of these mountains and extend well up on their sides. About the granitic rocks and immediately above them there is a bed of quartzites of the Lower Silurian age. It has an average thickness of about 150 feet. This sheet is not homogeneous, but consists of numerous layers of quartzite, some shales and some calcareous rock. The lower half is by far the more pure. Conformably upon these there follows a bed of impure siliceous dolomite, and then, ending the Silurian formation, a thin though generally persistent stratum of quartzite. This stratum is both thin and irregular. Its thickness varies from 10 to 40 feet. Emmons thinks this irregularity in thickness may be due to erosion, especially as he was unable to find any Devonian formation at this point. This stratum was the only one not identified on either the Morning or Evening Star, and, as the White limestone was found to be much thinner here than the average, erosion may have removed all of the quartzite and a part of the limestone at this place.

The Carboniferous strata are the most important, as it is in these that all the ore of Carbonate Hill and almost all that of the rest of Leadville is found. They consist first of a layer of blue dolomite, very pure when unaltered, but in the neighborhood of the deposits very much stained and changed. This is the stratum in which the ore occurs. The rest of the beds of this formation form a series of coarse grits and sandstones, with a few narrow layers of dolomite. When least eroded they are 2500 feet thick, but in the vicinity of the deposits now worked they have been almost entirely washed away.

Although places have been found where the various sedimentary formations succeed each other without interruption, such occurrences are extremely rare. There is scarcely a spot where, at some horizon, igneous rocks have not forced themselves in between the strata. These igneous rocks are acidic. They are regular quartz-porphyries or felsites, and are called by Emmons, and generally throughout the camp, porphyries, and that is the name that will be given them here. Many varieties may be recognized. White and Gray porphyry, of which more hereafter, are the varieties most intimately connected with the ore. These rocks are, comparatively speaking, seldom found in ordinary dikes, but occur for the most part as intrusive sheets, following the bedding of the formations in which they occur. Regular overflows are never found, and if they had once existed erosion would long ago have removed them. The most important sheets of porphyry occur above the Blue limestone, or in it. The larger occur above it. In thickness they average about the same as the Blue limestone, though in some cases they are much thicker, and on Carbonate Hill the White porphyry almost certainly attains a thickness of 1000 feet.

Besides this great complication, due to the intrusive masses of porphyry, the geology of the district has been rendered additionally complex by a series of faults which break the continuity of the beds. That these were made after the eruption of the porphyries is proved by the latter being broken too. Like evidence proves that the faults occurred subsequent to the deposition of the mineral matter. In direction the principal of these faults have a trend north and south, parallel to the mountain-range, though numerous minor faults run off from these at various angles. There are five or six of the larger ones. The western wall has fallen on all but one of them. Only two of these faults have any connection with Carbonate Hill; the Carbonate fault, running along the western slope of the hill, and the Iron fault, far to the east and dividing Carbonate from Iron Hill. The deposits were first discovered on the exposures made by erosion along these faults or the anticlinal folds in which they terminate; for here the superincumbent rock has been worn away and the ore itself, or the *iron* so indicative of ore, brought to the surface. The deposits thus outcrop like enormous contact veins, and some of the early locators took their claims parallel to the outcrop, supposing they could follow the ore with the pitch for an indefinite distance, a right denied them by the local courts, on the ground that the ore is not continuous from the outcrop down.

POSITION OF THE DEPOSITS.

The principal mines of Leadville are situated on Fryer, Iron and Carbonate Hills, all three of which form a group together at the foot of the Mosquito or Park Range of mountains. The distance from the summit of Carbonate Hill to Fryer Hill is about one mile to the north, and from the same point to the top of Iron Hill is three quarters of a mile south-east. By far the larger portion of the ore of the camp is derived from these three hills. Fryer Hill has so far produced the most, and Iron Hill next. It is Carbonate Hill that principally interests us.

Carbonate Hill lies south-east of the town of Leadville, and rises directly from its outskirts. Its summit rises to a height of over 10,600 feet above the sea-level, and over 400 feet above the main street of Leadville. To the east it slopes gently to the valley which divides it from Iron Hill. To the south it slopes into California Gulch, and to the north into Stray Horse Gulch, which divides it from Yankee and Fryer Hills. The claims which now have shafts sunk are situated along the western, south-western and north-western slope of the hill, and the oldest, including the Carbonate, Glass-Pendary, Ætna, Catalpa, Evening Star, Morning Star and Henriette, are along the outcrop of the ore-horizon, which approximately follows a contour about 200 feet below the summit of the hill. This outcrop, unlike the others, is covered with very little *wash*, much less than on Fryer Hill, yet it was late in being discovered. The distance of this outcrop to the Iron fault is about three quarters of a mile.

The Evening and Morning Star mines lie side by side on the north-western slope of the hill. The former lies to the south-west of the latter. The Blue limestone outcrops on the lower quarter of each claim. The Evening Star, though one of the best, is one of the smallest mines in the camp. It is a little less than half a claim in size, being 1305 feet long by 163 feet wide, or less than five acres in area. Its bonanza was so thick and covered so much of its area that it has afforded an enormous quantity of ore. Its location was in fact an extremely fortunate one, for it cut the great ore-body of this part of the hill through its thickest and richest part. If the nature of the body had been exactly known a better belt of this width could not have been chosen across it. To the south the ore begins to thin into the Catalpa, and to the north into the Morning Star, and it never again approaches in either of these mines the thickness it attained in the Evening Star.

The Morning Star mine lies next to the Evening Star. It is a consolidation of a number of smaller claims, of which the old Morning Star and Waterloo claims are the most important. It extends in width to the Henriette line, a distance of about 502 feet. Its eastern boundary is continuous with that of the Evening Star, but it extends further down the hill than the latter claim. The lower part of this property, however, is of little or no value, and there are now no workings run below 1300 feet from the east end. The relative size of the claims may be seen on Plate II. In both mines the ore occurs immediately below porphyry and in Blue limestone, as will be later fully described.

GEOLOGICAL STRUCTURE AT THE MINES.

The Evening Star mine is principally worked through four shafts. A fifth, now unused, is near the Forsaken shaft. A sixth, No. 9, Plate IV, has never had a drift run from it. It has, however, greatly aided in a thorough understanding of the formation, for it has been sunk nearly 200 feet into the Blue limestone horizon, and from this point a diamond-drill boring has exposed the succeeding strata for 200 feet more, and has thus given a key to the nature of the underlying formations. The Morning Star is worked through four shafts on its own property, and a fifth on the Evening Star ground, sunk conjointly by the two companies and used alternately by them. Three other shafts have been occasionally used but are now idle, and numerous prospecting shafts, or rather pits, also aid in studying the formation. The fact that the Morning Star is a consolidation of several claims explains the existence of so many shafts. The new McHarg shaft, now being sunk, was started October 1882, and is the only one that the present company has started. These numerous shafts and the underground workings which connect certain of them together have given a complete key to the geological structure. The data from which the conclusions are deduced are given, and accompanying these are two cross-sections, one through each mine. One of these sections, Plate III, is through the Morning Star, and is made along the dip of the beds. The shafts are projected on this plane and their depths given. A line of levels was run over the surface to determine the relative heights of the shafts at the top. The Evening Star is such a narrow claim that all the shafts lie nearly in one plane in that section. The same number indicates a particular shaft on the plan, Plate II, and on the sections, Plates III and IV. The numbers indicating the various shafts are as follows:

1. Upper Waterloo shaft.
2. Morning Star main shaft.
3. Boarding-house shaft.
4. Discovery shaft.
5. Old Waterloo shaft.
6. Lower Waterloo shaft.
7. Forsaken shaft.

THE MORNING STAR SECTION.

8. Old Forsaken shaft.
9. Evening Star No. 5 shaft.
10. Evening Star main shaft.
11. Evening Star upper shaft.

THE MORNING STAR SECTION.

The deepest shaft of this section is No. 1, the Upper Waterloo. It is the one furthest up the hill. This shaft begins in the White porphyry and passes through it for a distance of 360 feet. At this point the contact between the latter rock and the Blue limestone is reached, and fine ore is immediately struck. The second level, not shown in the section, is 45 feet below this point, and from here a drift runs to the end of the property 120 feet distant, where contact is again reached. All this region shows large bodies of fine lead-ore, which is indicated by pen-dotting on both sections wherever it occurs. Below the Upper Waterloo the Morning Star main shaft, No. 2 on the section, is the next. The surface at this point is 33 feet lower than at No. 1. The shaft, always measuring from the collar, which in this case is a distance above the surface, is 265 feet deep to contact. From here a main track incline, not shown in section, runs immediately below the porphyry and along the contact to the end of the claim, a distance of 480 feet. This incline is connected with the Upper Waterloo workings in several places, and the contact is always regular and void of all signs of disturbance. The Boarding-house shaft, or what Emmons calls the Lower Morning Star, is about 680 feet below the Upper Waterloo. It is shown on the plates as No. 3. This shaft passes through 75 feet of White porphyry to contact. It is connected with the shafts above by workings which show the porphyry to be continuous all the way along, and undisturbed by any break. No. 4 on the section, which is merely a prospect-shaft, sinks only 38 feet to contact, and here an incline runs for some distance, between 100 and 200 feet, and shows a regular contact.

If we connect the points where these shafts pierce the surface and those where they strike the contact we will have a section of the White porphyry, and by producing these lines down the hill till they meet we have the point where the White porphyry ceases. Immediately below the Blue limestone outcrops, or rather what was once the Blue limestone, for it is here for the most part replaced by vein matter, chiefly oxides of manganese and iron. This outcrop, in spite of the wash being in considerable thickness, is very plainly marked by the numerous boulders of iron-ore which are scattered in a broad belt along the face of the hill. It curves around the hill, running almost due east across the Henriette, and following the contour of the hill quite strictly. Numerous pits sunk by early prospectors along this zone show nothing but the iron-ore on the dump.

OCCURRENCE OF THE ORES.

The lower shafts belong to a different series, as they develop the ore under a second sheet of porphyry. They are many in number, but occupy little space in the section, as they all start near the outcrop of the second sheet. Most of the shafts have struck valuable ore here, and the Henriette has all her ore at this horizon. The Forsaken shaft, No. 7, Plates II, III, and IV, sinks 80 feet to contact, and 25 feet further to a second level, from which a drift runs to contact. Owing to there having been no record kept of the ground passed through in sinking this or any of the other shafts, it was difficult, and in one case impossible, to note the sequence of the layers or sheets passed through by them. In this shaft and in the Old Waterloo this was only obtained by climbing the ladders and looking between the round timbers whenever these were wide enough apart. It was especially difficult to do this in the Forsaken shaft, as the walls were generally some distance from the cribs and the intervening space was filled with wooden blocks. At the top there was only wash, boulders of iron-ore mixed with water-worn pieces of porphyry. More solid iron appeared below this and continued very nearly to the porphyry. Here a zone of impure though undoubted limestone occurred, which could be traced for 4 or 5 feet above the porphyry. Small pieces pried out showed the rock to be highly impregnated and discolored by metallic oxides, though it effervesced freely with warm dilute acid. At 35 feet from the surface Gray porphyry was entered, and this rock continued to the first level, 46 feet below. The layers passed through by this shaft were then as follows: Wash, 15 feet; iron-ore in place, from 15 to 30 feet; limestone in place, from 30 to 35 feet; porphyry, from 35 to 81 feet.

Below the Forsaken this sheet of Gray porphyry outcrops, as shown by the Old Forsaken shaft, No. 8, Plate III, which begins in that rock, and strikes the contact and ore at only 25 feet below the surface. The Forsaken workings follow the ore right to the outcrop, where the ore and wash mix with each other. The Portland shaft, not shown in the section, is a little further down the hill and does not touch the porphyry, but sinks down through iron-ore far below the ore-horizon and into barren ground.

The Old Waterloo shaft, No. 5 in the section, affords the best view of the ground along the zone below the White porphyry. This shaft, which is 135 feet deep, strikes the contact at 129 feet. There is but one level, from which two drifts are run; one horizontal, following the strike of the porphyry, and one an incline which follows the dip. The latter is shown on the section. The same iron-boulders and wash, merging into iron-ore in place, was passed through as in the Forsaken. At a depth of 63 feet limestone appeared, which soon became solid and quite pure, and which continued for 18 feet. At the top of this limestone layer there were seams of yellowish and much decomposed rock which resembled precisely some of the limestone found below the porphyry. Lower down the rock became massive and as pure as any encountered below the porphyries either in the Morning or Evening Star. The color of the more pure was hardly changed from the normal blue. The bedding-planes were plainly marked and corresponded in dip to all the sedimentary formations. The Gray porphyry began below this limestone. The contact between the two was finely shown, and the dip of the porphyry sheet was markedly different from that of the limestone layers, being much steeper. The existence of this limestone in place above the porphyry is alone sufficient, if there were no other proofs, to show that this sheet is a different one from the main sheet above.

As to the Lower Waterloo shaft, it could not be examined, so that the thickness at that point is not known. A rough record kept of the ground passed through in sinking the Evening Star shaft No. 5 (No. 9, Plate IV) shows it to have there had a thickness of 50 feet.

It will be seen on referring to the section that the Gray poryhyry has a much steeper dip than the White. This dip, though it has been maintained as far as development has gone beneath it, may lessen further down. Thus far it has cut across the limestone bedding to a lower horizon. Its course below the point where development has been carried is unknown. The course it probably takes is indicated in the section, this part of the outlines of the sheet being put in in broken lines. The Gray porphyry has been followed over 400 feet from the Lower Waterloo shaft. This distance takes it far under the White porphyry sheet. In fact, directly over the line of this incline, and only 205 feet from the shaft, there is an old shaft but 20 feet deep, which, though the bottom is filled up, shows walls of White porphyry, solid and apparently in place. Consequently the Gray porphyry has been traced at least 200 feet beneath the White.

The evidence of a second sheet of porphyry of no great thickness existing below the main sheet is therefore indisputable. This fact is recognized by Emmons, who mentions that the Half-way House and Henriette claims are on it, but owing to the undeveloped condition of the ground at the time of his survey he did not think that the Forsaken was under the same. He considered the Forsaken and the Lower Evening Star shaft (No. 9, Plate IV) to be under White porphyry, but cut off from the rocks above by a fault (the Carbonate fault) whose western wall had fallen. This fault he thought also ran above the Waterloo shafts. Recent workings have shown the Forsaken and Lower Waterloo to be under the same sheet of porphyry, for they are connected, and pay-ore runs from one shaft to the other almost without a break. Moreover, it has been shown that the White porphyry is broken at no point from the upper end of the claim to the point where it ceases, and that the Gray porphyry runs beneath it without a break for at least 200 feet. Consequently there can be no fault between the upper and the lower shafts. Acknowledging the existence of a second sheet of porphyry alone, there can be no doubt of this, for then there is no possible room for the displacement such a fault would cause.

The Gray porphyry goes unbroken to its outcrop. Below this no ore has ever been taken out. Some of the shafts below this outcrop have been sunk quite deep, but none could be visited as work was abandoned on them. Porphyry was certainly never struck in any of them. The Niles and Augusta, just below the Evening Star, was sunk all the way through the White limestone and into the upper strata of the Cambrian quartzites, so that these lower formations must come quite near the surface at this point. No shaft along the foot of the hill strikes White porphyry. The fact that the fault does not lie above the outcrop of the second sheet of porphyry by no means proves that one does not exist below. If, however, it does exist below, the displacement must be much greater than that inferred by Emmons (170 feet). It seems more likely that the outcrop of the Blue limestone marks the crest of an anticlinal fold which the beds make at this point, and which has been eroded until the White porphyry cap has been worn completely away.

The section through the Evening Star, Plate IV, has been used in this description as well as the other section, Plate III. It shows little that has not already been mentioned. In this section the Gray porphyry is seen to send up a wedge-shaped dike into the limestone, which almost reaches the White porphyry. Such dikes, breaking up from the lower sheet, occur elsewhere.

To sum up, the following are the important points regarding the ore-bearing sheets:

(1) All the formations dip towards the south-east (E. 25° to 30° S.). To this there is no exception. At the same time the surface rises towards the south-east.

(2) The upper sheet of White porphyry is not continuous over the face of the hill, but the Blue limestone outcrops below it.

(3) A second sheet of porphyry different from the first lies below it. The intervening space of about 175 feet is taken up by vein-matter or Blue limestone.

(4) No porphyry sheet of any kind has yet been developed, and no ore found below the outcrop of this second sheet of porphyry.

(5) The formations from the eastern end of the properties to the outcrop of the second sheet of porphyry (Gray porphyry) are undisturbed by fault or fold.

The formation below the Gray porphyry is well shown by the No. 5 shaft of the Evening Star (No. 9 on Plate IV), or rather by the core of a diamond-drill boring made from the bottom of it. A meagre record preserved of the ground passed through in sinking this shaft gives only the depth at which Gray porphyry was struck and the thickness of the sheet. As the shaft was not in use and had no ladders it could not be examined, but the dump showed only iron, Gray porphyry and Blue limestone, but no parting quartzite. The drill-core also showed no

THE WHITE LIMESTONE.

trace of the latter stratum, but White limestone began immediately. This rock continued for a distance of only 100 feet, and then the sheet of Cambrian quartzite came in and the rest of the boring was in this formation. There was no way of determining the exact boundary between the Blue and the White limestone, and it is accordingly put in in broken line. It will also be seen that at this point the White limestone has not its average thickness (about 150 feet according to Emmons).

DETAILED DESCRIPTION OF THE VARIOUS ROCKS.

The various formations and their order of occurrence having been described, there remains to be given some idea of their nature and characteristics before leaving them and passing to the especial consideration of the ore-deposits and vein-material in the Blue limestone horizon. The different rocks will be taken in order, beginning at the bottom and going towards the top.

THE CAMBRIAN QUARTZITES.—This formation, though not entirely passed through by the drill, was, in all probability, nearly pierced, for the description of the strata given in Emmons's report closely tallies with the section here exposed, and would lead one to infer this. As it is, the lower 100 feet of the boring is in this formation. At the bottom of the boring the rock was very hard. No large pieces of core came up, but generally coarse powder and small fragments of very hard quartzite of dull gray color. A few feet further up the ground was more impure and contained much ferruginous material. Numerous assays, made by Mr. Bonner, the assayer at the Evening Star mine, show that this material carried silver. One sample gave him $14\frac{1}{2}$ ounces, and another 5 ounces; all the rest ran much lower. An assay of material very near that which ran $14\frac{1}{2}$ ounces gave the writer only $1\frac{1}{2}$ ounces. This impure streak ceased 8 feet above the bottom of the boring, and very pure, white, saccharoidal quartzite came in and continued with but one break for 60 feet. The break was caused by an impure streak 10 or 12 feet thick near the bottom. At 70 feet from the bottom the quartzite began to be impure. Red, iron-stained, calcareous quartzite came in which at times showed micaceous specks. The succeeding space of 25 or 30 feet marks a gradual change into the White limestone, and is occupied by layers of impure rock sometimes containing little calcareous matter and sometimes a great deal. Some of these layers are shaly, and one streak of very fine-grained argillaceous shale, soft, slaty and of even texture, occupied a zone of 8 or 10 feet just at the top of these quartzites. This stratum comes in mixed with almost pure quartzite, and goes out mixed with impure limestone.

THE WHITE LIMESTONE.—The White limestone comes in above the quartzite, but the change is so gradual that no point can be taken as the boundary between the two. It is most impure and siliceous at the bottom, but is characterized throughout by a high percentage of silica. It contains very little organic matter. The color is a dirty white or grayish, very different from the characteristic blue of the Blue limestone. An analysis of this rock is given a little further on. The drill-core showed very little vein-matter in this rock, only a few streaks of wad filling crevices. These occurred near the upper part of the bed.

THE BLUE LIMESTONE.—Although this rock contains all the ore, and though all the workings are situated in its horizon, the drifts seldom directly encounter the rock, as they run where the ore and other vein-matter occur. The limestone has, however, been struck in some places both on the Evening and Morning Star, and especially in the southern workings of the Forsaken and Lower Waterloo. It is always stained and altered in color, and more or less highly impregnated with iron and manganese. Some of the shafts which have been sunk deep into the formation have struck more pure rock, notably the upper Evening Star, at 100 feet below the White porphyry. An old quarry on the side of Iron Hill, towards California Gulch, from which this rock used to be taken for fluxing at the smelters, shows a face of 35 or 40 feet of very pure Blue limestone. Here it is massive and deep gray-blue in color. It shows the bedding-planes very perfectly. The only thing that breaks its homogeneity is the narrow streaks of calc-spar which have frequently separated along the bedding-planes. The rock splits easily along the bedding-planes, so that flat plates may be split off six inches or more across.

An analysis shows the Blue limestone to be a very pure dolomite. The rock in the mine resembles the rock exposed in the quarry in every way except purity and homogeneity of color. The specimen already referred to from the upper Evening Star shaft was just like the quarry rock, even in color, but was the only perfectly pure specimen found in either mine. Wherever found elsewhere it was always stained brown by manganese and iron, and altered in composition by the addition of these ingredients and of silica. The bedding is always easily recognizable even when in a very much altered state. Below are given three limestone analyses. No. 1 is of the White limestone; No. 2 is of a pure specimen of Blue limestone from the quarry on Iron Hill; No. 3 is of a very much altered specimen of Blue limestone from near the White porphyry in the Upper Waterloo.

	1	2	3
Silica	22.71	1.05	6.87
Iron sesquioxide and alumina	1.18	0.25	1.38
Manganese dioxide	....	....	2.43
Lime	24.26	30.37	28.64
Magnesia	15.48	21.15	17.43
Carbonic acid	36.26	46.80	43.07
Sulphur trioxide	.07	trace	.05
Phosphorus pentoxide	.10	"	.04
Organic matter	trace	.40	not determined
	100.06	100.05	99.91

THE GRAY PORPHYRY.

The very high percentage of silica in the White limestone is characteristic of that rock. The very low percentage of silica in No. 2. is prominent. No. 3 has probably derived much of its silica from the porphyry, which is but a few feet from it. This analysis may represent the composition of the deeply stained rock occurring elsewhere in the mines.

The jointed structure in this stained rock is always distinct, and it is a very significant fact that wherever found, whether as isolated "boulders" in vein-matter or in large bodies, the dip is always the same. The ore and vein-material found at this horizon will form the subject of the following chapters.

THE GRAY PORPHYRY.—This rock, where exposed to view by the workings, lies entirely in the Blue limestone and in the lower part of that formation. As the sections show, it corresponds in dip very closely to the enclosing rock for a distance of 150 to 200 feet from the outcrop, but from a point just above the Lower Waterloo shaft it takes a much steeper dip and cuts down across the limestone stratum to a lower horizon. Probably it again takes the dip of the formation lower down (Emmons). It is not at all improbable that originally this sheet continued to rise across the Blue limestone until it reached the White porphyry, and that the point of juncture is now entirely worn away by erosion. The steeper dip is, on the whole, very regular from the Lower Waterloo shaft as far down as the incline has been run, but there are many minor local irregularities in the pitch which often seem to bear some relation to the deposition of ore. These will be mentioned later on.

The popular name for this rock is Bird's-eye porphyry, on account of its spotted appearance, due to the separation of feldspar crystals from the ground-mass. It differs very much from the White porphyry, even when both exist in a very decomposed condition side by side. This sheet of rock is so thin, being only from 46 to 50 feet in thickness, and has been so subjected to decomposing agencies, that an approximately fresh specimen could nowhere be obtained. Consequently no analysis of the rock was made, as it was thought that one would give no idea of the original composition and nature of the rock. Even at the centre of the sheet, where the firmest specimens of the rock were found, kaolinization, which is the peculiar form of decomposition of this rock, had reached an advanced stage, and all the specimens were softened and crumbled when dry. The feldspar-crystal outlines could be recognized. They occur with great frequency, ordinarily as small grains a sixth to an eighth of an inch in diameter. Very rarely the outlines of much larger crystals an inch or so in length could be recognized. They are always soft and almost completely changed to clay.

The quartz exists as small, vitreous, transparent, white crystals about as large as the smaller feldspar crystals. They are not numerous, but any small lump weighing an ounce or so would be likely to show a few of these granules shining through it.

The ground-mass is generally of a dull gray color and harder than the rest.

The feldspar crystals appear as granular white dots through it and occupy about one half the surface. Sometimes the ground-mass has a decided pinkish color, which is probably due to the oxidation of the protoxide, which is said to stain the fresh rock greenish.

It is not an uncommon thing for branches to run from this sheet into the formation above, and even to extend as far up as the White porphyry and to flatten out against that rock. Such a dike occurs in the Evening Star. The main shaft, No. 10, Plate IV, penetrates it, and shows that it comes almost to the White porphyry. This dike is shown in the section. Another dike, not shown in the sections, occurs in the Morning Star. This body rises all the way up to and flattens out against the White porphyry and cuts off the ore-body in that direction. Another such dike occurred in the workings of the Boarding-house shaft. All of these dikes show all the characteristics of the sheet below, and have also decomposed in a like manner.

Immediately along the lower contact of the Gray porphyry, except where undecomposed limestone comes in direct contact with it, the rock has been completely changed to clay which is soft and perfectly plastic and devoid of all signs of the original texture. Further from the contact the rock, still very soft, begins to show the characteristic mottled appearance, and further in yet, generally two or three feet, it becomes firmer and the jointed structure appears. Though very prominent, the jointing is not carried as far as in the White porphyry, but only splits the rock into large lumps.

THE WHITE PORPHYRY.—The White porphyry is a more regular sheet than the Gray, and seems to follow the dip of the Blue limestone with great regularity. The facts concerning its position have already been stated. The Brooklyn mine, lying east of the Maid of Erin (a claim adjoining the Morning Star on the east), finished a shaft through this rock last fall, so that a section through 500 feet was exposed on the dump and as fresh specimens as occur were obtainable. The rock is light gray in color and has a fine-grained, even surface on which only very few porphyritic ingredients can be distinguished. The jointed structure is exceedingly prominent and is carried very far, so that the rock, especially if at all decomposed, breaks into small angular blocks. The jointing faces are always smooth and sometimes have an almost perfect polish. They commonly have stains of black oxide of manganese which are often beautifully dendritic and give rise to the popular name of forest-rock. Besides this light homogeneous-looking rock a peculiarly banded variety is found, of a light rusty or reddish-brown color which is due to stains of sesquioxide of iron. A little of this variety occurred in this shaft. In other places on the hill it is very abundant. Except in the banded appearance, which is only due to the coloring matter, it does not seem to differ very much from the rest.

The macroscopic crystals are scattered very thinly through the fine-grained base. QUARTZ occurs in irregular crystalline granules, not more and generally less than 0.1 inch in diameter. They are colorless, semi-transparent and vitreous, just as those of the Gray porphyry, but they occur much less frequently than in the latter rock. MICA is of more frequent occurrence than quartz. Some of it exists as black mica or biotite, and some of it as the white mica or muscovite, the latter being a product of the decomposition of the other constituents. The biotite is in larger pieces than the muscovite, but is less frequent. In some cases it is much decomposed. The muscovite is fresher than the biotite, and frequently shows hexagonal sections very nicely. The FELDSPAR crystals were very seldom recognizable in the ground-mass. Very rarely a small, shining crystal of this mineral could be seen which resembled sanidine. Orthoclase is the essential feldspar of the rock. Many cavities now empty or only partially filled with oxide of iron were observed in this rock, and though none of them contained any of the original mineral which once filled them, it is probable that it was iron pyrites, all of whose sulphur has been completely washed away.

When a thin slide of this rock was examined under the microscope it was found to be a true felsite or quartz porphyry, though it has been much altered by secondary changes. The feldspar has become, as a rule, opaque from kaolinization. The ground-mass was fine-grained, and a high power had to be used to examine it. The porphyritic ingredients were quartz and monoclinic feldspar, the former predominating. Between or about the larger crystals of quartz the ground-mass was sometimes twisted in currents with a peculiar feathery arrangement of the small crystals which strongly suggested the fluidal texture. In another slide of the banded variety mentioned above the decomposition had proceeded much further, and the characteristic features could not be seen nearly so perfectly. In both sections the ground-mass was feldspathic, but contained much free quartz as tiny angular specks. On comparing the section of the freshest specimen with a number of slides of normal quartz porphyries from Europe it was found to have a precisely similar structure, though the porphyritic ingredients were generally larger and more prominent in the foreign specimens.

The following is an analysis from as fresh a specimen of White porphyry as could be obtained, yet it has evidently lost some alkali:

Silica	74.98
Iron sesquioxide	1.27
Alumina	15.27
Manganese dioxide	1.07
Lime	1.03
Magnesia	trace
Phosphorus pentoxide	"
Sulphur trioxide	"
Soda	1.89
Potash	2.10
Water	2.00
	99.61

It is true that the White porphyry decomposes and gives rise to large bodies of clay, but the whole rock does not soften in this manner as the Gray does. Commonly it decomposes to a dry siliceous mass in which the jointing is perfectly distinct and which crumbles to small bits along those planes. Such decomposed rock caves easily and has to be supported with stout timbers. Ground of this kind is most damaging to timbers, not so much because it swells and crushes them, as because it is so loose that it caves and a great weight of rock is let down on them. Where the rock decomposes to a soft clay-like product this caving does not occur to such an extent, and the drifts do not have to be so stoutly timbered. Thus beneath the Gray porphyry in the Lower Waterloo the timbers need not be nearly so strong nor so close together as under the White porphyry.

THE ORE-DEPOSITS.

THE ORE-CURRENTS.

The position of the ore in the formations is shown in the sections, Plates III and IV, by pen-dotting. It occurs in the horizon of the Blue limestone and immediately under porphyry. In some deposits about Leadville the ore occurs in White limestone and other formations, but by far the larger part of the output of the whole camp comes from the Blue limestone, as in the Morning and Evening Star mines. The contact between the limestone and the porphyry is also the normal position of the ore, especially on Carbonate Hill, but there are important exceptions to this rule in the camp, notably in the deposits of the Silver Wave-Cord group on Iron Hill, in which the ore occurs as lenticular bodies in limestone and in no way connected with the porphyry. On the Morning Star a large amount of ore has been found beneath the Gray porphyry sheet, as already mentioned. This body differs in no essential way from that underlying the White porphyry, the ores being similar and the laws governing their occurrence the same.

Speaking roughly, we may say that the Blue-limestone bed is now occupied by two classes of matter, (1) the original limestone, and (2) vein-matter of various kinds which has supplanted the original rock. Of these the latter class lies immediately under the porphyry. The limestone is below. The lower surface of the porphyry is comparatively even, and there is, as a rule, an abrupt boundary between it and the vein-material. Between the limestone and the vein-matter the boundary is very irregular. Sometimes the vein-matter is so thick that it almost supplants the limestone, and again so thin that the limestone comes very near the porphyry, and even in direct contact with it, pinching out the mineral matter entirely. Thus, on going to the lowest level of the Evening Star main shaft, No. 10, Plate IV, and following the drift to the point where it emerges from the Gray porphyry, limestone is found, which, however, is deeply stained and discolored, while numerous heavy seams of hydrous oxides of iron and manganese run along the bedding-planes. It is, in fact, so altered that there can be no doubt that vein-matter proper would appear only a short distance above, though the ground just above could not be examined. The south-western workings of the upper shaft of the same mine, No. 11, Plate IV, show a quite pure limestone rising up to contact and cutting off the vein-matter altogether.

Though the vein-matter occurs along the contact, it by no means occupies all or nearly all of it. Also, it does not occur in isolated patches which have no relation to one another. It is found to follow well-defined courses or channels in the limestone and along the contact. These courses are much longer than broad and, though they vary in both width and thickness, are perfectly continuous and uninterrupted. Such streaks have been excellently named ORE-CURRENTS.

All the ore is found in these currents. The larger part of the ore of the Morning Star and almost all that of the Evening Star has come from one great current under the White porphyry, the only one yet developed under this rock on these claims. This current begins on the western slope of the hill in the Crescent claim. It starts from the outcrop and passes from here in a north-easterly direction through the Catalpa and into the Evening Star. In the Crescent the thickness of the ore alone is not very great, but it becomes thicker and broader in the Catalpa, and passes into the Evening Star a splendid body which rapidly develops till an enormous thickness is reached, the maximum being 70 to 80 feet. The ore in the current as rapidly decreases in thickness towards the Morning Star, and the current passes on much thinner but broader and still bearing fine bodies of ore. It passes through the south-eastern part of the Morning Star consolidated property into the Maid of Erin and Big Chief claims. The Brooklyn shaft, beyond the Maid of Erin, also strikes ore and, as it lies in the right direction, is almost certainly on the same current.

Below the Gray porphyry there is another current which, though not nearly so large as the preceding, contains very fine bodies of rich ore. This current, as the plan, Plate II, will show, has two branches which run into the hill and, joining, go on together as one current. One of these branches, the smaller, starts at the south-western corner of the Evening Star from below the outcrop of the porphyry and runs in a north-easterly direction toward the Old Waterloo shaft (No. 6, Plate II). The other, larger branch begins at the outcrop on the old Half-way House claim and runs a little south of east into the Lower Waterloo and Henriette, the boundary between these two mines being about the centre of the current. Continuing in this course it is joined by the smaller branch. This ore-current has lasted as far as developments have been pushed, and there is every reason to suppose that it will continue into the hill. It has already been followed from the outcrop to a depth of 400 feet below the surface, and the breast of the main incline is already over 250 feet beyond the point where the outcrop of the White porphyry ceases.

The pen-shading on Plate II and the pen-dotting, Plates III and IV, represent approximately the extent and depth of the ore alone in these currents. It would be more difficult to give the boundaries of the entire current, gangue as well as ore, as the gangue is not followed when there is no promise of ore. If the entire amount of vein-material were represented on the plates, the axis of the current would not be changed. The current would only appear broader and thicker. Of the area represented in shading on Plate II, at least two thirds or three fourths is occupied by pay-ore. No ore occurs outside of these limits.

Outside of such a current the limestone comes to contact. Under the White porphyry it is not often met with. In the Morning Star it was found in only two places, and there it was altered and impregnated with impurities. The Boarding-house shaft showed a great deal. It was black and contained so much iron that it became strongly magnetic on testing with the blowpipe. The other point was in the face of drift 59 of the Upper Waterloo, where the same characteristics prevailed except that manganese was relatively more abundant. These two points indicate the boundary of the upper current on the north-east. The limestone of the upper Evening Star shaft (No. 11), which has already been mentioned, is on the southern side of the current.

Below the Gray porphyry the waste vein-matter does not extend far beyond the point where the ore ceases, and the boundary of the current is better defined. All along the southern side of the smaller branch the drifts have exposed quite pure limestone coming up to contact, once not more than 25 feet from where the ore ceased. Limestone is likewise found in the barren ground just above the junction of the two branches. The northern side of the main current has never been determined. It extends beyond the Henriette and perhaps outcrops along Stray Horse Gulch.

These are the only currents that have been developed on the property; and now the very important question arises, Do any others exist, and if so, where? It is simply impossible to say whether there are or are not other currents, and the only way of finding out is by sinking shafts or running expensive drifts through barren rock to the unexplored parts of the mines. As to where they will be found, if they do exist, of course the answer is easier. It is almost certain that no other body of ore underlies the White porphyry in the Evening Star, for the unexplored area in that mine is very small. In the Morning Star there is a large area north-west of the main current which, like the ground beyond it on the Henriette, is as yet undeveloped. This area will be easily explored from the new McHarg shaft which is being sunk to the current under the Gray porphyry. The probabilities seem to be against there being another current in this ground; for if there is, some of the old prospecting-shafts lower on the hill should have struck it. The drifts that will be run from the new shaft, which is already sunk 150 feet below the contact (May, 1883), will solve this question.

The greatest unexplored area is on the second contact; that is, under the Gray porphyry. It is the entire area south-east of the current laid open under that body. This ground will soon be explored, for both the middle and upper shaft of the Evening Star (Nos. 10 and 11) are being sunk to explore it. If an ore-current does exist there, drifts run from these shafts will find it. In the mean time the results of exploration have to be awaited.

THE VEIN-MATERIAL.

The vein-material as it now exists consists of carbonate, sulphide and sulphate of lead, sulphate of iron, oxides of iron and manganese, and silica. The latter is sometimes in a very pure state as spongy deposits from solution or as chert, but more often is mixed with greatly varying proportions of oxide of iron. There is abundant evidence to show that the vein-materials originally existed as sulphides and oxides (or carbonates), but owing to the decomposition of the former by the meteoric agencies to which they have for ages been subjected the other, secondary products have arisen. In the original deposits the minerals forming them occurred pure in considerable quantity, but the larger part of the vein-material consisted of mixtures of certain of them in almost every proportion. Consequently, after the further changes due to the action of meteoric waters, the variety of the materials going to make up the vein-matter is very great. The number of pure minerals, however, which occur in the deposits is very small. The following were noted:

CERUSSITE, lead carbonate, $PbCO_3$.

GALENITE, lead sulphide, PbS.

ANGLESITE, lead sulphate, $PbSO_4$; Pyromorphite, lead phosphate and chloride, $3Pb_3P_2O_8+PbCl_2$.

CERARGYRITE, horn-silver, containing chloride, bromide and iodide of silver.

CALCITE, calc-spar; BARITE, heavy spar, barium sulphate, $BaSO_4$; Calamine, silicate of zinc; Native silver; Pyrite; Rhodochrosite, carbonate of manganese;

Psilomelane, hydrous manganese dioxide; Dechenite, vanadate of lead and zinc, (Pb,Zn) V_2O_4; BASIC SULPHATE OF IRON.

Of these minerals the ones in small capitals only are of common occurrence. Besides the well-defined minerals there are many mixtures of manganese and iron oxides; iron oxide and silica; silica, iron oxide and carbonate of lead, etc., which exist in very large quantity and form the major part of all vein-material.

The vein-matter may be most conveniently if not naturally divided into ore and gangue. Concerning the position of these, it may be said that the ores occupy almost identically the same relation to the gangue that the whole of the vein-matter does to the limestone. The ore lies immediately under the porphyry, and the gangue below it. The ore does not occupy the whole surface of the current, though it forms a continuous sheet. It is almost always found in strength along the centre, but the gangue always rises to contact along the sides. The thickness also varies greatly. The rule that where the ore is very thick the gangue below is likewise so, and vice versa, does not hold at all, although there is always a large amount of gangue beneath the ore.

THE GANGUE.

IRON-ORES.—These consist for the most part of hydrous oxides of iron and manganese. The lower half or three quarters of the current is largely composed of these materials. Further up towards the top large quantities of siliceous material occur, especially in the upper current. These mixtures of oxides of iron and manganese are both compact and loose. The loose is found most frequently near the top of the zone of gangue rather than at the centre. It also occurs near the bottom, and frequently forms seams in the limestone and about boulders of that rock. It is a soft wad-like material, but does not powder. It can be easily dug down with a pick and falls in flakes. It contains some sulphuric acid which has most probably been derived from the oxidizing galena. In color it is from dark brown to black, and is slightly mottled.

The rest of the lower part of the current is occupied by the firmer variety which forms the greater part of the vein-material. It is of a deep blackish-brown color, solid and hard. It has a well-marked jointed structure, and in many ways has, when in large masses, the outward appearance of the stained limestone. In both currents the iron is in such a high state of oxidation that the surveyor uses his compass freely in the mines. All the manganese exists as the dioxide.

The following analysis represents the average composition of this material. The specimen was taken from the Lower Waterloo.

Silica	10.73
Iron sesquioxide	46.22
Alumina	0.06
Manganese dioxide	31.18
Lime	1.20
Magnesia	0.68
Water	9.98
Phosphorus pentoxide	0.05
Sulphur trioxide	0.03
Carbonic acid	0.54
	100.67

THE SILICEOUS GANGUE.—This exists in large bodies and in great variety. It occurs, as a rule, near the ores and both above and below them. Low down in the current it is rare. When above the ore and along the contact it is very pure. It is then often a spongy, white, amorphous material, with hard, cherty cores, and has evidently been deposited from solution. In the Lower Waterloo and Forsaken there is a large sheet of compact rock, now very much disintegrated, but showing a jointed structure and many external features of quartzite. This sheet overlies all the ore of the branch current. It will be mentioned later.

The siliceous gangue proper belongs below the ore, and occurs in large though ill-defined and by no means regular or continuous bodies. It is much more common under the White porphyry than under the Gray. It is never pure, but has always a high percentage of sesquioxide of iron. In its purest form it exists in local concretionary deposits of a red jasper imbedded in more impure material. It is then much harder and more compact than the rest and has few joints running through it. As long as the percentage of silica is very high the rock is extremely hard, but as the siliceous character becomes less prominent and oxides of iron and carbonate of lead appear in quantity it assumes a regular jointed structure, so that a blow of a hammer will shatter the brittle rock into small angular blocks. The most common variety of this gangue has from 30 to 50 per cent of silica, and the rest is made up of sesquioxide of iron, carbonate of lead and a little water. An analysis of this kind of material is given below. The specimen is very rich in silver, but that metal came almost entirely from coatings of cerargyrite deposited from solution along the joints already alluded to.

Silica	39.50
Silver chloride	1.44
Lead oxide	15.46
Iron sesquioxide	35.67
Lime	0.50
Carbonic acid	3.44
Water	3.46
	99.47

The percentage of lead carbonate in the analysis given above is seen to be quite large. A full series of this kind of rock running from a quite low percentage in lead, like the one above, to a quite high percentage of that metal occurs in the Morning Star. Very little of it comes under the Gray porphyry, though it is even then occasionally met with. In both the Evening and Morning Star it is abundant under the White porphyry. Where the percentage of lead is low the rock is generally not perfectly homogeneous, but some parts of it contain more lead than others.

CLAYS.—The small quantity of alumina in the vein-material is noticeable. It exists almost exclusively in clays which may be traced to porphyry as a source. All the largest sheets—and here they reach considerable proportions—lie directly along it, and the very small quantity existing below is in most cases plainly due to infiltration. Along the upper contact clays are, comparatively speaking, scarce. For the most part the White porphyry at contact does not kaolinize, but assumes a hard, dry, siliceous form which in a yet more advanced stage of decomposition easily crumbles to a dry, sandy powder and shows no clay at all. In the upper shaft of the Evening Star and in the Boarding-house shaft there were considerable bodies of clay along the contact. These sheets were highly impure and contained many streaks of manganese oxide and some carbonates of lime and manganese.

Clay is common along the lower contact. The Gray porphyry is soft and tends to change to clay, and generally several inches, at least, of very pure clay lies above the vein-matter proper. Often it is peculiarly striped and stained by iron and manganese. It is very pure and much of it is of surprising whiteness. It seldom contains appreciable quantities of sulphur. That all these clays have arisen from the decomposition of the porphyry in place there can be no doubt, for there is always a gradual change from the pure structureless clay without to undoubted porphyry within. The thickness of these clay sheets varies from a few inches to two or three feet.

The remarkable clay called by the miner "Chinese talc" has a different mode of origin. This material comes in the mines both along the contact and below it. The black iron of the Lower Waterloo shows many pockets of it which are entirely isolated and from five to ten feet from the contact. Some of it is stained by manganese and iron, but most of it is pure white. It has a conchoidal fracture and an opalescent or pearly lustre, and is semitransparent on thin edges. On exposure to the air it slowly becomes opaque, especially if impure. Some specimens from the Carbonate mine which were soft became opaque on a very short exposure. Two specimens from the Lower Waterloo were analyzed. No. 1 was the softer. It was stained light green when taken out of the mine. On six months' exposure it became opaque on the surface and had turned to a pinkish color. Within it was still fresh and green. It contained only a faint trace of manganese. No. 2 was much harder and more brittle, and remained translucent on thin edges even after half a year's exposure. The following are the analyses:

	(1)	(2)
Silica	42.94	40.53
Alumina	37.11	38.51
Water	19.48	19.43
Sulphur trioxide	0.65	0.33
Lime	0.66	1.50
	100.84	100.30

These analyses show that the composition is variable, though No. I gives the formula $Al_2Si_2O_8 3H_2O$ very nicely.

A few minerals which occur in the deposits, most often in the gangue, may be briefly mentioned here.

CALCITE.—This is of course common; generally it occurs as the ordinary incrustation filling crevices in the waste or lining cavities in the iron. In the large cavities in the gangue beautiful crystallizations are often found. Large and very handsome rhombohedrons on cerussite occur in the Evening Star. Much of it occurred in a cave found in the workings of the Boarding-house shaft.

BARITE.—Heavy spar in small quantities is also very common. Generally it is found scattered through ore as small crystalline aggregates of a pure variety. A more impure kind, of pinkish color, is found in larger masses.

CALAMINE.—This mineral is far more rare than either of the above. It is almost always found in waste, most frequently in the Lower Waterloo, forming the filling of crevices and druses. It occurs in aggregates of fine, slender, needle-like prisms.

DECHENITE.—Dechenite has been found in very small quantity in the Evening Star. It occurs as an incrustation on a siliceous gangue. When thick these coatings are of a deep brick-red color. Surfaces six inches across have been found completely covered with it. It yields no reaction whatever for chlorine, and gives fine reactions for vanadium, lead and zinc before the blowpipe.

THE ORES.

There can be no doubt that the lead was all originally deposited in the limestone horizon as the sulphide, and that the silver was either deposited along with it as the sulphide, as in ordinary argentiferous galena, or perhaps also with this, but separated out as argentite. Besides these there seems to have been deposited in some places a great deal of iron pyrites, either mixed with varying proportions of galena or nearly free from it. The oxidation of these sulphides has proceeded so far that, comparatively speaking, very little galena is now left, and there is not a trace of iron pyrites. The products arising from this decomposition are for the most part carbonate of lead, chlor-bromide of silver (containing also some iodide), sulphate of lead and sulphate of iron. The original sulphides formed a belt along the contact. In most places the change to the oxidized products was a gradual one in place. Small amounts of all the metals were carried off in solution by the oxidizing waters. They were to a large extent redeposited in the gangue, and in some cases sufficiently concentrated to form ore out of what would have otherwise been too poor to pay for extraction.

The galenas occupied the upper portion of the belt of sulphides, and the pyrites, where it occurred in any quantity, formed a belt below. The galena has changed to carbonate of lead, and the pyrites to hydrous basic sulphate of iron. Any galena that was mixed with the pyrites has likewise changed to the sulphate.

The three sections given (Plate I) illustrate the position of the ore in the deposits and represent the different zones of ore and gangue.

Section 1 is through a portion of the ore-body of the Upper Waterloo. The galena has entirely changed to the carbonate, and the pyrites to the sulphate of iron. The former is represented by the gray-blue band immediately under the White porphyry, and the latter by the yellow band below the blue. The latter zone was not rich enough to pass as ore, and its lower boundary could not be accurately determined, as it was seldom exposed. It is, however, very nearly the same as that given in the section. All the other lines on the plate are from actual measurements. The boundary between the carbonate and sulphate is abrupt and well defined, as is that between the latter and the iron-ore lying below it.

Section 2 is through a part of the small branch current of the Lower Waterloo. The same description holds here as in the former case. The blue belt is very highly oxidized lead-ore, rich in lead but containing some silica, and by no means as pure as the preceding. The yellow zone, which in this case is fine ore, occupies the same position as in the other section. In this section the current is not very deep, and the stained, impure limestone, is indicated in the lower part by the conventional lining.

Section 3 is also from the Lower Waterloo, but is from the main ore-current below the junction with the branch. This body of ore never had a zone of pyrites below it, and hence there is no yellow sulphate. The cerussite, which is highly ferruginous, still contains many nodules of galena. The iron immediately below is somewhat soft and contains much manganese dioxide and some sulphuric acid. It soon changes into the firmer variety.

No section of the great ore-body of the Evening Star is given, as it was for the most part stoped away, and there was no way of getting an accurate section representing how and as what the ore occurred.

THE GALENA.—There is not much of the galena left in the ore-bodies, though in one or two places it is abundant. It occurs nowhere as continuous and solid bodies. It everywhere shows signs of an advanced state of decomposition into oxidized products, and is divided into nodules by seams or streaks of these. It rarely occurs under the White porphyry, and then in small nodules imbedded in the carbonate of lead, and seldom over a few inches in diameter. It is most abundant in certain parts of the Lower Waterloo. No galena is found in the small branch current (Plate II), as oxidation has there been very complete, but in the larger branch and the main current, especially in the former, it is quite abundant.

The ore of the Half-way House mine, below the Lower Waterloo, consisted mostly of this mineral, and was very rich, although the bodies were not large. About the first level of the Lower Waterloo shaft (No. 6, Plate II), galena occurs as nodules and broken streaks surrounded by carbonate of lead. Below this point, for 50 to 100 feet, it is more abundant than at any other place in either mine. Here in some places it is in streaks, alternating with seams of carbonate, and together with this forming a very rich pay-streak two or three feet in thickness. The galena streaks are bright at the centre but lustreless on the outside. They are not solid, but in several main streaks and numerous branches which form a network. The decomposition products are deeply iron-stained to a rusty color. The galena in decomposing first loses its metallic color and forms a dark band of greasy lustre, half an inch in breadth ; then, while the streak is still hard, the brown color appears. At the centre of the broadest of the streaks of carbonate it is softer and granular. The galena in this body was on the whole coarse-grained, but almost every texture could be seen even in different parts of one and the same seam of undecomposed mineral.

About at the same level as this, but over on the Henriette line, galena is found in great abundance and with very little carbonate of lead. Here it is along the contact, but it does not form a solid streak. It occurs as a series of streaks of the greatest irregularity running through and lying in black iron. In some cases it lies directly along the porphyry, and in some it is entirely surrounded by iron. The streaks are always near the porphyry, however, and the zone in which they occur never extends more than six or eight feet from it. These streaks are small as well as extremely irregular, but rich in both silver and lead. The iron-ore about it contains no lead and far too little silver to pay. Assays of this iron gave 6 to 20 ounces of silver. The decomposition of the galena was slight, but narrow streaks of cerussite ran through the seams and divided it up into nodules which were both bright and large, many of them weighing 50 or 100 pounds. The iron in which they occurred was somewhat soft. It contained much manganese dioxide and some sulphuric acid.

Below the junction of the two currents galena is more rare. It is only found as nodules, and these are mostly small and imbedded in large bodies of carbonate of lead. On breaking them open the centres are always bright. Around this core there is a zone of the lustreless sulphide, and then this changes into carbonate so gradually that a division line cannot be recognized between the two. The transition of the galena into cerussite is always plainly shown.

In the above examples there is every step from nearly solid galena in which the change is just beginning, to the almost completely oxidized bodies in which there is scarcely any galena left. In the branch current of this contact, and also in the Upper Waterloo, we have bodies in which the changes have long been completed. The change is an altogether gradual one from without inward, and there can be little doubt as to the manner in which it is effected. The oxygen dissolved in the waters penetrates the ore and oxidizes the galena to sulphate of lead. The calcium carbonate, held in solution in the water by an excess of carbonic acid, acts upon the sulphate of lead, forming carbonate of lead and sulphate of calcium. The latter passes off in solution. Alkali carbonates would act on the sulphate of lead in the same manner.

There are often cavities in the galena, and these are generally lined with large transparent crystals of cerussite. They are then in long prisms capped by the pyramid.

Galena has been found on several occasions within the Gray porphyry. Sometimes it forms the filling of irregular gashes which run up into the rock from the contact, but which are always very small. On two occasions it was found forming seamlets in it four or five feet above the contact, and running parallel to it. In one case the seam was over a foot thick and mostly carbonate of lead, with galena in large nodules. In the other the seams were only six inches thick, but were nearly solid galena.

The galena is generally very rich in silver. Under the White porphyry samples were too rare to afford trustworthy estimates of the general run of the original galena of the upper current, but under the Gray porphyry there was an abundance. To give a good idea of the richness of these galenas fifteen specimens from various parts of the Lower Waterloo were assayed. The average silver contents was 180.2 ounces per ton of 2000 pounds. The different samples varied greatly in value, the minimum being but 36 ounces and the maximum 490. Four of the fifteen ran less than 75 ounces, and three of them over 300. A small nodule found later gave by assay 1142 ounces, but this was far above any other sample of galena found in either mine. The percentage of silver seems to be in no way connected with the texture of the galena or its crystalline structure. Almost all the galena, coarse or fine, is rich, but both coarse and fine do occur which are, comparatively speaking, very poor in silver. The galena in the Lower Waterloo is not as a rule very pure. Only two of the fifteen assays mentioned above gave over 75 per cent of lead. Nine of them gave over 70 per cent. The average of the fifteen was about 69 per cent. In the Upper Waterloo the galena is very pure and, as far as could be judged, less rich in silver. The impurities are commonly iron oxide and silica, chiefly the latter. The iron is generally invisible. Only one piece showed tiny specks of pyrites, the only traces of this mineral found among the ores in either mine.

THE PURE CARBONATE OF LEAD.—This is and has always been the chief ore of both mines and under both sheets of porphyry. The purest, largest and best-defined bodies of carbonate of lead occur under the White porphyry. In the Evening Star they had been removed before the fall of 1882, but in the Morning Star they were well developed but not stoped out. In the upper levels of the Forsaken and Lower Waterloo there is another body of large extent, though neither so thick nor so pure, but very completely oxidized. The ores of the lower levels and elsewhere are still less pure, and contain the galenas already mentioned as well as iron oxide and other gangue. The purest carbonate of lead bodies contain little else than pure, granular cerussite crystals, often so loosely put together that they easily crumble between the fingers, especially when damp. Some of it is more firmly cemented and, especially when dry, quite firm and hard. The grains out of which the purest bodies are formed are about the size of coarse sand and resemble it superficially. They consist of very imperfect crystalline granules, among which, by the aid of a strong magnifier, a more or less perfect prism may occasionally be recognized. They are of course very brittle, and, whatever the color of the body be, the powder is white. It is for this reason that a pick always leaves a white mark in this ore.

In the color of the pure cerussite ore there is great variety. Rarely it is an almost pure white. Brown, grayish brown and gray are the most frequent colors, but grayish blue is also common. Sometimes it is a light cream-color, sometimes almost black. The coloring-matter is mostly oxide of iron, more rarely oxide of manganese, and only forms a small percentage in the composition of the ore. The color bears no relation to the silver contents, and rarely to the percentage of lead. The bodies of the Upper Waterloo frequently show a peculiar banded structure, very faint and delicate, and entirely in the coloring-matter. These bands can be plainly seen by candle-light, but can hardly be recognized on the surface. They consist of bands of slightly different color, and, though gently wavy, always run strictly parallel to the dip of the body. Almost all the purest cerussite ore shows this banding.

The percentage of lead in these ores is almost that of normal cerussite. Two analyses were made, one from the ore of the Upper and one from that of the Lower Waterloo. The first was grayish white in color, the second bluish gray.

	(1)	(2)
Gangue	0.42	4.43
Lead oxide	83.27	77.70
Silver chloride	0.03	0.12
Lime	....	0.80
Carbonic acid	16.30	14.56
Sulphur trioxide	0.32	0.28
Phosphorus pentoxide	....	1.22
Chlorine	....	trace
Moisture	....	0.50
	100.34	99.61

These two analyses represent the purest ore found under the White and Gray porphyries respectively. The first is almost pure cerussite. The second shows, besides gangue, a considerable percentage of pyromorphite (nearly 8 per cent).

These pure carbonates are very fine ores on account of the high percentage of lead; for, since they are very desirable to the smelter, a high price per pound is paid for the lead in them, and a low charge deducted for treatment. They are, however, poor in silver. Wherever the ore is purest and oxidation has been most complete the amount of silver is found to be the least. In the Upper Waterloo the pure cerussite ore with only a few per cent of gangue normally runs not more than 20 ounces in silver, and on an average even less. A good idea may be gained of the amount of silver in these high-grade lead-ores by referring to the ore sales. From 19 shipments (or between 900 and 1000 tons) of ore running over 50 per cent of lead by assay, the average was but 15.03 ounces per ton, or 0.0516 per cent, of silver. When this ore is rich it occurs in pockets in the poor bodies, and the silver is always visible as grains and scales of the chlor-bromide.

THE IMPURE CARBONATE OF LEAD.—The pure heavy carbonate of lead ore just described forms a large part of the ore of both contacts. The more impure ores of carbonate of lead are likewise abundant. These may consist of a mixture of pure carbonate of lead with a visible gangue, yet too finely mixed to admit of their being stoped separately. As examples of this class we have streaks of cerussite alternating with narrow streaks of iron oxide or sulphate, or some siliceous gangue, or seams of it running up into and mixing with the porphyry. But besides these, which differ in no way from the large bodies, there are the HARD CARBONATES of the miner. This name has a rather broad significance, and may be used to indicate any sort of hard rock that carries silica and iron sesquioxide, with enough carbonate of lead and silver to make it pay as ore.

In describing the siliceous gangue it was shown that lead occurs in that material. Whenever the percentage of lead is high, 20 to 30 per cent, we have a hard carbonate. A perfect series of such material is found, from the hard jasper-like rock with little or no lead, on the one hand, to the pure lead carbonates, on the other. As the name implies, these ores are hard. The ones containing much iron and silica are very hard and scratch glass easily. As the percentage of lead increases they become softer, but are always hard as long as they contain any appreciable amount of silica. Excepting the poorest of them they are perfectly homogeneous in texture and uniform in color. The color varies through almost every shade of brown, red and gray.

Hard carbonates do not occur in extensive sheets along the contact like other ores, but rather as thick, massive bodies of small area which underlie pure carbonate and extend down into waste. When the percentage of lead is quite high, say 40 per cent or more, there are frequently streaks or veinlets of pure, soft cerussite running through it. The richest variety of this class of ore—that is, the richest in lead—is of a light gray color and contains little iron. It is extremely solid and heavy and will contain from 80 to 90 per cent of carbonate of lead, the remainder being mainly silica. Another very pure variety has very little silica but much iron. This variety is deep red in color and not nearly as hard as the more siliceous. The composition of such an ore is given below in analysis No. 1. For the other analysis I am indebted to Dr. LeRoy W. McCay of the Scientific School, Princeton. It shows the composition of a fair average specimen of hard carbonate in all but the silver. The latter metal is present far in excess of the average.

	(1)	(2)
Silica	2.82	18.84
Silver chloride	0.16	1.30
Lead oxide	66.98	54.89
Iron sesquioxide	14.23	11.38
Lime	trace	1.78
Carbonic acid	13.07	10.83
Sulphur trioxide	0.55	....
Water	1.52	1.61
	99.33	100.63

No. 2 is deep brownish-red in color and very hard. It has a perfect jointed structure which divides the rock into small rectangular blocks. Though this jointed structure is very common, it is seldom as perfect as this. It is on these joints that the cerargyrite which makes the ore so rich occurs.

Several thousand tons of such ores have been taken out by the present management, much of which was stored in old cribs and drifts, and the rest in place through the old works of the former company. In some cases almost the entire filling of an old drift would be shipped as ore, after passing through the hands of the ore-sorters, and would realize as much as $20 to $30 a ton above all expenses.

The average amount of silver in these ores is much larger in proportion to the amount of lead present than it is in the pure cerussites. It is seldom that a hard carbonate assaying 30 or 40 per cent of lead carries less than as many ounces of silver. A large amount of the silver ore has often been deposited along the crevices in it by percolating waters. Such is generally visible in scales.

The question now arises as to the source of the carbonate of lead in these siliceous ores. That the water dissolves some lead and redeposits it below admits of no doubt, for the incrustations of cerussite along joints and cleavage planes are very common. But it cannot be supposed that this solution and redeposition has been so extensive as to account for much more than a trace of the lead in the hard carbonates. The only way they can have arisen is by the oxidation of impure galenas. A perfect and complete series of these ores, from impure ones not assaying over 20 per cent to the purest gray varieties running 60 to 65 per cent lead, was collected, all of which contain bright particles of galena undergoing decomposition. In all of these the various stages between the undecomposed galena and the thoroughly oxidized carbonate may be seen. It cannot be doubted that these hard carbonates have arisen from impure masses of galena just as the more pure carbonates of lead have arisen from bodies of more pure galena.

THE BASIC IRON SULPHATE.—The most peculiar material in the mines is the hydrous basic ferric sulphate which so frequently underlies the lead-ore in both the upper and lower workings. Though this substance is only sometimes very fine ore, it is grouped with the ores because it is often rich and always contains some lead and silver. It forms sheets which, with remarkable persistence, extend beneath the cerussite ores in certain parts of the mines. Its position is shown in sections 1 and 2, Plate I, and is represented by the yellow color. In color it is a pure light-yellow ochre and has a great resemblance to yellow clay. It is never plastic except when wet. Usually it is dry and firm and has a perfect jointed structure, due probably to pressure. It consists of iron sesquioxide combined with sulphuric acid, has much water and varying amounts of sulphate of lead. When subjected to intense heat it loses all its water and most of its sulphuric acid combined with the iron. The following is an analysis of a specimen of this material:

Lead sulphate	29.18
Silica	trace
Iron sesquioxide	40.22
Alumina	1.11
Sulphur trioxide	18.02
Water	11.20
Silver chloride	0.27
	100.00

This sample contains more lead sulphate than the average. The percentage of lead being low, the value of this material depends largely on the amount of silver it contains. In the Upper Waterloo large bodies of it occur under the pure lead ore which contain on an average only a few ounces, and are therefore too poor to mine. In the upper contact this substance is rarely rich, except when it contains chloride of silver in visible form. In the Lower Waterloo and Forsaken it is abundant in certain places, chiefly under the cerussite ore of the branch current. Here it always runs well in silver (from 20 to 100 ounces), and makes an excellent ore. The silver is never visible in this body. It has most probably arisen from the galena, which has given rise to the sulphate.

The fact that this material, wherever it occurs, is always perfectly pure, welldefined and abruptly bounded by the other vein-material indicates without a doubt that it has all arisen in the same way, and differently from any of the rest of the vein-material about it. Although the soft black iron which often lies below both this and the other ores contains some sulphuric acid, this is not often in large quantity and some of the iron contains none. Moreover, the iron contains no lead and, even when under very rich ore, not enough silver to pay.

The manner in which it is believed to have been formed has already been stated; namely, by the oxidation of a belt of pyrites mixed with more or less galena (seldom over 15 or 20 per cent) which formerly occupied this position and formed a layer with the more pure galena above it. This easily oxidized mineral, subjected for a great length of time to agencies which have almost completely oxidized the last traces of galena in the ore above it, has completely succumbed and been brought to its highest state of oxidation. Not a trace of pyrites has ever been seen in it. Small nodules of galena of great richness in silver have occasionally been found in it. These nodules assay from 60 to 70 per cent in lead and average as high as 450 to 500 oz. in silver. They are seldom entirely bright at the centre, but even there have begun to decompose along the cleavage planes.

THE CHLORIDE OF SILVER.—The great difference in silver contents between the carbonate of lead and the galena is very striking. The amount of silver in a thoroughly oxidized carbonate of lead is never normally above 40 or 50 ounces. Even when no richer than this the silver may often be seen as tiny grains and scales of chloride scattered among the granules. The average richness of the carbonate of lead has already been shown to be much less than this.

The galena, on the other hand, is (as has already been stated) much richer and averages at least the half of one per cent silver. From ten samples of ore assayed, five of bright large galena nodules and five of the thoroughly oxidized carbonate of lead around them, it was found that in proportion to the lead present in each there was over six times as much silver in the galenas as in the cerussites.

It is easy to explain where the silver missing from the carbonates has gone. All the iron about the ore-bodies contains silver. All the porphyry along the contact contains it. And it is not merely the material about the ore. The limestone and the vein-matter deep down in the current often contain traces of silver.

Assays of these materials yield from one to four ounces. Scarcely any vein-matter can be obtained that does not run one or two ounces, and much of it runs higher. In some places the amount of silver deposited is much more considerable and then ores arise, it may be out of what before was far too poor to be such. In these ores the silver-bearing mineral can always be recognized. In both mines it is invariably the same in color, light greenish yellow. It is soft and sectile, and feels like lead between the teeth. Exposed to the light it does not change color in the least. Qualitative tests show it to contain chlorine, bromine and iodine, all in considerable quantity. It occurs as scales or plates and as single crystalline grains, or aggregates of such grains, and as rough crystalline coatings on the walls of crevices and joints in the various vein-materials; also as scales and grains more or less thickly scattered through granular carbonate of lead, and as highly crystalline lumps forming the lining of druses and hollows in that mineral, just as cerussite crystallizes in the hollow galena nodules. Though generally highly crystalline, the grains are small and have to be examined with a magnifying-glass.

The most common ore formed by the deposition of cerargyrite is low-grade and siliceous. It is the hard, siliceous gangue, with numerous joints and crevices which, interlacing and extending all through the rock, allow the mine-waters to trickle through them or stand in them. The chloride is deposited along these joints. As long as this is visible it is stoped down and sent to the surface, where it is carefully sorted. Of course this is not done when the base itself is pay-ore. Chloride seldom enriches the iron sufficiently to make it pay. When the yellow sulphate of iron is firm and has crevices running through it, it also frequently shows chloride of silver. The pure carbonates of lead frequently have it, and it has occasionally been found in lumps of decomposing galena.

The most remarkable deposit of ore arising from the deposition of cerargyrite occurs in the White porphyry on the Evening Star. This body was struck by the main shaft between 20 and 30 feet above the contact. The gangue is nothing but porphyry. This is in a state of extreme decomposition here, and there is a deep stain of iron oxide along all the joints which divide the rock into lumps of but a few inches in diameter. The silver is mostly invisible, but shows here and there through the stained portion as tiny specks of chloride. The amount of lead present is very low and it is often altogether wanting. This body is very irregular, but in many places it is 8 or 10 feet thick and three or four times as broad. The ore could only be distinguished from the waste by assay. Just below this point a large dike of Gray porphyry has broken up and reaches almost to the White. It seems not improbable that the disturbances accompanying this eruption may have shattered the White porphyry and allowed the solutions bearing the metals a limited access to this rock.

Rich pockets of ore always owe their richness to the chloride of silver that has been concentrated there. They generally occur in the pure carbonate of lead. Specimens of the latter metal are frequently found, which run from 5 to 10 per cent of silver, but there is seldom over a few pounds of such ore found at one place. Lumps of this cerargyrite weighing a few ounces have frequently been met with. Only one or two lumps weighing over a pound have ever been found. Such pockets are always small, and it is rarely that many tons of ore averaging over 100 ounces are found at one place.

THE ORES IN POSITION.

There has been so much ore removed from the current under the White porphyry that it is difficult to give a good idea of the ore in the current taken as a whole. Its thickness in different parts and its richness may be shown from the size of the stopes and their thickness, from the appearance of the streaks of ore found in the old workings and by referring to the ore sales, but with regard to the ores forming the different parts of the body, their relative size and their position very little information can be given. Under the Gray porphyry as clear a view as could be desired is obtained of the ores standing, in all parts of the current.

THE UPPER CURRENT.

This current reached its greatest thickness in the Evening Star. The thickness of the ore at the centre of the current was, as already stated, very great. At the centre and north-western side the ore varied from 30 to 60 feet in thickness and contained so little waste that there was often great trouble in getting material with which to fill the cribs. From the centre the ore rapidly thinned towards the south-eastern side, till at the upper shaft it was very thin, and a little beyond it ceased altogether. The bottom of the ore-body was extremely irregular and ended in a series of branchlets, streaks and isolated bodies running through or lying in the gangue below. Much of the ore at the bottom was very siliceous, hard carbonate of lead; some of it low-grade ore consisting of iron impregnated with cerargyrite. The top of the current was more regular, being against the porphyry. It must not be thought that the porphyry was a perfectly smooth sheet. On the contrary, it was very irregular, but the irregularities were all local. The sheet taken as a whole is smooth enough, but it has innumerable hollows and gashes, many of them of considerable size. These were generally filled with ore or silica.

The quality of the Evening Star ores is shown by the sales to have been very fine. The percentage of lead is low when compared with the Morning Star ores;

but in silver they run much higher. In Part II., a table is given showing the production of ore for both mines and their average richness for August, September and October 1882. This table shows the stated differences very plainly. The average of the Evening Star for the year before would have been considerably higher in both silver and lead. The silver for that time averaged from 50 to 60 oz. per ton; the lead was still far lower than the average in the Morning Star portion of the same current, which is also higher in lead than the average of the Lower Waterloo ores. In the Evening Star the top of the main body was richer than the bottom, especially in lead.

The ore-body thinned as it approached the Morning Star boundary, and entered that claim a much thinner body, though the area occupied by ore was just as large. In some places along the centre it was still very thick, sometimes 20 or 30 feet. At other places along the sides of the current some barren spots occurred, but the continuity of the ore was never broken. The percentage of lead in the ores begins to increase and the silver, though still high, begins to decrease in amount. The ground in this part of the property was stoped by the old Company; but the nature of the ores can be inferred from the streaks overlooked which are every now and then discovered. There seems to have been along the contact some very pure carbonate of lead which ran very well in silver, from 25 to 40 oz. The contact was very irregular and had along it much spongy silica, some of which was quite pure and some mixed with ore. Below the pure streak of "sand" carbonate there was much very highly siliceous ore containing scarcely a trace of iron. The ore passed on towards the Upper Waterloo shaft, No. 1, Plate II, where there was very pure cerussite along the contact and a great deal of hard carbonate, now often containing iron, below. More of this hard ore occurred in this part of the property than any where else in either mine. Beyond this the rest of the current was still standing last fall. The hard carbonates cease and streaks of basic sulphate of iron begin to appear under the ore just beyond the last-mentioned shaft. The pay-streak was weak at first, but rapidly strengthened and developed into a fine body of large area of which section 1, Plate I, is typical. The blue of this section shows the lead-ore, which assays from 68 to 75 per cent lead and averages as high as 70 to 72 per cent. It lies immediately below the porphyry. The section shows neither the thinnest nor the thickest part of the body, but represents a fair average. The yellow band represents the yellow sulphate of iron. This material contains, intimately mixed with it, from 10 to 30 per cent of anglesite, and sometimes a little carbonate of lead. In silver it generally runs from 4 to 6 oz. In section 1 this material is seen to reach the contact at one place, cutting off the cerussite. It runs on beyond this point for 30 or 40 feet, maintaining a thickness of 10 to 14 feet, and then thins out and disappears, ordinary iron taking its place. These sulphates, being of so low a grade, are left in the mine as waste, or if they have to be removed they are sent up for the dressing dump.

This body of ore, though of so low a grade in silver, is remarkable for the rich pockets that occur in it. These are both large and small. The small ones often contain little nuggets of chloride of silver. There are only one or two of the larger pockets, and they do not contain excessively rich ore. They are from 20 to 30 feet through each way, and vary from 50 to 300 ounces in silver, with an average of 75 to 100 ounces. The yellow sulphate below these pockets is also pay-ore. Whether this rich ore has arisen from a concentration of the silver from the poorer ores around it, or whether it comes from galena richer than the rest, it is hard to decide. The fact that in the Lower Waterloo galena samples taken from the same drift, and from what was once the same body, differ greatly in their silver contents shows that the last is certainly possible.

Before leaving this current it may be worth while to describe a dike of Gray porphyry which breaks up in this part of the mine on the northern side of the current. This dike rises to the contact and spreads out along it. The ore thins as it approaches it and ends, at some places, in direct contact with it. The only drift that cuts through this dike shows it to rise up with a thickness of 8 or 10 feet at this point. It lies in a sheet along the White porphyry above, in some places 30 or 40 feet wide. At the western end an upraise shows it to have a thickness of about 25 feet. No ores were found beyond this dike, but when the drift stopped there was limestone in the face of it. The dike showed the characteristic structure of the Gray porphyry. It was in a highly decomposed condition, soft and changing into clay just as the decomposition goes on in the lower workings. All about it the White porphyry decomposes to the dry siliceous product, often showing specks of white mica which has been so frequently mentioned.

THE LOWER CURRENT.

The smaller branch current under the Gray porphyry begins at the lower north-west corner of the Evening Star claim, and runs with a north-easterly course till it joins the larger one. The area occupied by pay-ore in this current is remarkably large, though the ore is never very thick, being seldom more than six or less than two feet through.

In the Evening Star there is very little carbonate of lead. Almost all the ore is the basic sulphate of iron mixed with anglesite. The carbonate of lead is not continuous, but occurs in narrow pure streaks of small extent, generally along the contact and always underlaid by the other ore. Sometimes small lenticular strips of carbonate of lead occur imbedded in the sulphates but parallel to the contact. The thickness of the pay-ore in this part of the current is variable.

OCCURRENCE OF THE ORES.

Four feet is a good average. Along the sides of the current the ore is very irregular and ends in deep gashes running into the waste, from which it is easily distinguished by the color. Below the ore there is sometimes much ferro-siliceous gangue, very hard, but more commonly only the black iron-ore.

After passing the Morning Star line the ore in the current weakens somewhat and, along the lower south-eastern side especially, is very thin. At the same time the carbonate of lead becomes relatively more abundant, until at about a line drawn across the current through the Old Forsaken shaft, No. 8, Plate II, there is a continuous seam of "sand" over one of the yellow ore. Though here only a foot thick, the two bands are remarkably well separated from one another, as analysis No. (2), page 34, and that on page 36 show very well. The two specimens from which the analyses were made lay within six inches of each other in the mine, and were both taken from a specimen of only a few pounds' weight, of which the lower half was the sulphate and the upper the carbonate.

From this point the ore begins to thicken and rapidly develops into a fine body which extends without change almost to the Old Waterloo shaft. This body has a continuous streak of lead carbonate above it, and one of the yellow ore below. On the north-west side of the current the ore is the strongest all the way along, and the carbonate of lead streak is thinner than the sulphate, the ratio being about 1 : 2. On the lower south-eastern side the pay-streak is not so strong, and the carbonate of lead is relatively much thicker. Section 2, Plate I, shows a section of this body running along the centre of the current. The carbonate of lead streak is seen to be very much the thicker of the two here, but further up towards the north-western side this is not the case. .

Above this body of ore there is a siliceous sheet, soft, loose and resembling moist sand more than anything else. Further from the ore it is harder, and has a jointed structure which divides it into small, angular blocks but a few inches in thickness. This firmer material resembles a softened and disintegrated quartzite. It varies from 2 to 10 feet in thickness, and when penetrated soft Gray porphyry is found beyond. The lead carbonate mixes with this sandy material. At the bottom of the streak it is as a rule pure, as shown by the analysis, page 34. Towards the top it gradually becomes more impure and merges into this soft siliceous layer, so that in most cases no boundary can be distinguished.

Though averaging much lower in lead than the Upper Waterloo ores, the carbonate of lead is richer in silver in proportion to its lead contents. It assays from 15 to 30 ounces in silver and from 30 to 50 per cent of lead. Below this streak the yellow sulphate is very pure and always good ore. It is low in lead, but is richer in silver than the carbonate. Assays show it to vary greatly in value. From 20 to 80 ounces per ton in silver and from 1 to 20 per cent in lead are the usual amounts of these metals in this ore; the average is from 40 to 50 ounces and from 5 to 15 per cent. Most silver is found, as a rule, where there is most sulphate of lead, though this is not always the case. No specimen showing visible chloride was ever observed in this current.

This ore-body begins to diminish in breadth and thickness about 50 feet before the Old Waterloo is reached. The yellow ore diminishes most rapidly and opposite this shaft appears only in isolated patches, mostly below, sometimes above, the carbonate of lead. Up to this point the dip of the Gray porphyry has been quite gradual, but here the line is reached where the entire sheet assumes a much steeper dip (from 30° to 35°). The crest of this bend is marked by the Lower and Old Waterloo shafts, both of which are situated upon it. The ore passes over this bend and still continues a pay-streak, though much changed in nature. Both the siliceous sheet above and the sulphate below have disappeared, and the ore now consists of narrow streaks of granular carbonate of lead, with a great deal of highly siliceous and very hard ore below—so hard that it barely pays to mine it, although it yields $10 to $15 net per ton from the smelter.

The other branch current and the main current below their juncture are very different. The ores of the northern branch current are principally galenas. They do not occur in large, continuous bodies as do the ores in all other places, but form a series or string of small bodies which are, however, very rich. The ore in this part of the mine overlies black iron-ore except in the case of a single pocket of large size, where a seam of yellow ore, very thin but exceedingly rich, occurred. Above the lower shaft, No. 6, where the dip is not so great, the ore is in larger bodies than just after the steep dip is begun. At one place the ore occurred as small irregular seams of galena, seldom as much as a foot thick, scattered along the contact, and in black iron-ore through a zone reaching 6 to 8 feet from the porphyry. Everything has to be stoped in such a body, and the galena separated from the iron by careful sorting. The largest of these bodies always contains much carbonate of lead, as already sufficiently mentioned under the head of Galenas.

Below the juncture of the two currents the ore-bodies are larger and more completely oxidized. They also run much less in silver than the galenas. Section 3, Plate I, is taken along the strike through the ore in this part of the mine. It is at a point only a short distance below the juncture of the smaller current. It consists of a low-grade carbonate of lead of average quality in silver. The impurities, which are abundant, are mainly ferruginous. Galena occurs only as nodules, and now forms but a small portion of its bulk. This body begins about 75 feet from the Henriette line and extends southward to the other side of the current. Toward the east it stops suddenly on the crest of a very steep dip taken by the porphyry at an angle of at least 70°. It continues at this rate for about 20 feet, and then resumes its regular dip. The ore begins again 30 or 40 feet beyond, and has continued from this point on as far as exploration has been pushed, a distance of about 100 feet. The grade of the ore has improved a little also, and the average of silver is higher than has been struck anywhere else in the current except in the galena ores already mentioned.

The difference between the degree of oxidation of the two branches is very striking. The fact that the ores of the one are completely oxidized while those of the other are not nearly so, and that below the junction the ores, though further from the surface, are more completely oxidized than the ores in the large branch, shows undoubtedly that the smaller current has for some reason been the course of greater quantities of meteoric waters. As the smaller current cuts across the dip of the stratum and makes a sharp angle with the outcrop, all the waters entering at the outcrop and following the contact between the outcrop of the current and its junction with the main channel would go down to it, while the other current, being perpendicular to the outcrop and following the dip of the porphyry, would be the recipient of comparatively little. That water did pass along both currents many channels washed out along the contact prove.

THE DEPOSITION OF THE VEIN-MATERIALS.

There can be little doubt as to the way in which the vein-matter was deposited. The evidence tending to show that "*the process of deposition of the vein-materials was a chemical interchange, or actual replacement of the rock mass in which they were deposited,*" seems the more indisputable the more one sees of the deposits. The great changes due to extreme oxidation and the action of large quantities of surface-waters for such a long period of time render it impossible to trace step by step the way in which the replacement was accomplished, but that it did take place there is sufficient proof.

During September and October, 1882, the upper shaft of the Evening Star was sunk, from 8 feet below the contact, 100 feet towards the Gray porphyry sheet. It was mentioned above that this shaft strikes the south-eastern edge of the main upper current. The ore here was not very thick, much of the area had no pay-ore, but there was from 4 to 8 feet of vein-matter along the contact about the shaft. On sinking the shaft, limestone was entered at the very start; namely, at 8 feet below the porphyry. It was stained and much altered, coarse-grained but firm, although not nearly as solid as the normal blue limestone. This limestone continued only 10 feet, and at 18 feet below the contact iron was again entered. Both the upper and lower sides of the limestone layer were parallel to the dip of the formation, and the bedding-planes, which were well marked, were likewise parallel to it. The iron below was of the common soft variety and contained sulphates. It had no limestone in it. This streak lasted but 7 feet, and like all the others, above and below, was parallel to the dip of the formation. Limestone was again entered at 25 feet below the contact. This layer was so softened and disintegrated that it had lost all coherency and was very loose and crumbly. The bedding-planes were completely obliterated. It is met with in other parts of the mine, and the miners have the special name "lime-sand" for it. At 35 feet another layer of iron appeared, and continued without interruption for over 40 feet. This iron was firmer than the other layer, but still contained some sulphuric acid. At about 50 feet it contained from 5 to 20 per cent of lead, and here and there nodules of galena poor in silver. This lead bearing material occupied a zone parallel to the dip of the beds. At about 80 feet "boulders" of limestone began to appear, imbedded in the vein-matter. These isolated lumps showed, when large, the bedding-planes in position and having the regular dip. They became more and more frequent, until at 100 feet solid limestone was reached, the purest struck anywhere on either mine. At this depth the sinking was temporarily stopped.

It would be hard to account for these layers of limestone and "boulders" all in position if a pre-existing cave were assumed. They can only be satisfactorily accounted for by supposing the iron to be in the position of former layers of limestone, which allowed an easier passage for the flowing waters, and that along these courses the vein-matter was deposited by replacement.

All the iron which lies near the ores or along evident water-courses, although containing much more iron and manganese as oxides, has always a considerable though variable amount of basic sulphates. The yellow ore which underlies the lead carbonates in some places, and which is supposed to have been formerly iron pyrites with a little galena, consists, as has been already stated, altogether of basic sulphate. It would seem that if the conditions, when these sulphides were decomposing, favored the formation of basic sulphates, the latter compounds would have existed in large quantity in the iron lower down, provided this had formerly consisted of the sulphides of iron and manganese. The analysis given on page 27 shows that the more pure variety of iron is almost free from sulphur. The main mass of the vein-matter, in fact, contains little of this element. Further, although the iron often reaches a thickness of 60 to 100 feet, a trace of pyrites has in no case been discovered in it. It seems more probable, therefore, that the iron and manganese were deposited by the ordinary reactions between mineralized waters and limestone, the solution depositing its protoxides of iron and manganese and replacing them by calcium and magnesium oxides. The iron and manganese now forming the main body of the currents would accordingly have been deposited as the carbonates. From these the hydrous oxides as they now exist would be formed by the oxidizing waters. The silica must have been deposited by the replacement of limestone by that substance, molecule for molecule. That the lead and silver and some of the iron were deposited as the sulphides there can be no doubt. The proofs of this have, however, been given under the description of the different ores.

SOURCE OF THE VEIN-MATERIAL.—As to the source of the vein-material, no careful study of the question was attempted, as neither time nor data could be found. The White porphyry, of which an analysis is given on page 21, showed no traces of lead or silver, but also showed scarcely a trace of sulphur, though the rock at one time certainly contained much iron pyrites. Although it seems strange that such immense quantities of vein-matter could be derived from a sheet of porphyry, in this case not more than 1000 feet thick and on an average not more than 200 or 300 feet thick, it is hard to account for the formation of the currents in any other way. There is absolutely no proof, as far as could be ascertained, that the vein-matter came from below; and though development has been extensive throughout the camp, none of the alleged "feeders" have been struck.

The only theory regarding the formation of the deposits that is founded on a careful, thorough and protracted study of the deposits of the whole camp is that given by Emmons in his abstract of his main Government Report. It is only through such a protracted examination of the deposits, involving not only a careful inspection of all the mines but a great number of delicate chemical analyses, that any trustworthy conclusions can be reached. The data from which his theory is deduced are not yet published, but his published conclusions account most satisfactorily for all the phenomena of the deposits developed in the Morning and Evening Star mines.

PART SECOND.

METHODS OF EXTRACTING THE ORES.

THE SHAFTS.

There are two general methods of taking ore out of the mines in use at Leadville; one by means of an incline starting from the surface and running into the hill along the mineral zone, the other by means of a shaft sunk perpendicularly until the ore is struck. The latter method is used on the Morning and Evening Star properties, though often in combination with the former; that is, with an incline following the ore from the bottom of a shaft along the contact, through which the ore and waste are conveyed to the shaft by the hoisting-engine.

It has been stated in a preceding chapter that the Morning Star mine has four shafts at present supplied with engines, and that the Evening Star has likewise four, of which one is used by the two mines in common. The reason for such a number of shafts is obvious. The ore occurs in two separate bodies under different layers of porphyry, and, as far as present developments show, widely removed from each other horizontally. Hence the ore of each mine could only be taken out through one large shaft by running long and expensive drifts and by handling the ore a great number of times. It is also more convenient to have several shafts on each body of ore, because the very slight dip of the porphyry soon carries the workings a long way horizontally from the shaft, giving rise to the same difficulties mentioned above. On the other hand, the distance of the ore below the surface is so short, varying from 80 to 400 feet, that the cost of sinking is not very great, while a number of shafts allows less work for each, and they can be small and supplied with light machinery.

On the Morning Star all the shafts, with one exception, were sunk to contact before the consolidation, and that one was nearly completed, so that the present company has not, until lately, found another necessary. On the Evening Star the main shaft was already sunk to contact when the property was bought in 1879. The upper shaft was sunk later by the present company, and the Raworth shaft by the side of the main shaft when the latter proved inadequate for raising the ore of the main ore-body.

All these shafts are small. They are rectangular in shape and have two compartments, one for a ladder-way, the other for hoisting. In the Lower Waterloo the ladder-way has been converted into a pumping-compartment. In the new McHarg shaft of the Morning Star there will be a large pumping-compartment, but no ladder-way. The following tables will show the dimensions of the various shafts:

SHAFTS ON THE MORNING STAR.

Name of Shaft.	No.	Depth to Contact.	Dimensions of Ladder-way.	Dimensions of Hoisting-Compartment.	Size of Shaft in the clear.
Upper Waterloo	1	360	4 x 5	5 x 5	5 x 9.2
Main Shaft	2	265	3.8 x 4	4 x 4	4 x 8
Old Waterloo	5	135	3.4 x 3.5	3.5 x 3.5	3.5 x 7
Lower Waterloo	6	140	3.3 x 3.3	3.3 x 3.3	3.3 x 7
McHarg Shaft			4.5 x 5	4 x 5	5 x 9

SHAFTS ON THE EVENING STAR.

Name of Shaft.	No.	Depth to Contact.	Dimensions of Ladder-way.	Dimensions of Hoisting-Compartment.	Size of Shaft in the clear.
Main Shaft	10	116	3.7 x 4	4 x 4	4 x 8
Upper Shaft	11	300	3.2 x 3.2	3.2 x 3.3	3.2 x 7
Raworth			3.7 x 4	4 x 4	4 x 8
Forsaken	7	81	3.2 x 3.3	3.2 x 3.6	3.2 x 7

COST OF SINKING THE SHAFTS.—There were no records kept of the cost of sinking the old shafts. The new McHarg shaft has reached a depth of 300 feet (May 1, 1883), of which White porphyry occupied the first 180 feet, and iron and limestone the remaining distance. The cost of sinking this 300 feet, including putting the timbers in place, was not quite $18 per foot. The cost of the timber and framing brings the total cost of the shaft completed to $23.50 per foot.

METHODS OF TIMBERING THE SHAFTS.—The soft nature of the ground, and its tendency to swell or cave, render substantial timbering necessary; and even when the larger timbers are used they often bend in and distort the lining. The ordinary methods of timbering shafts, which are in use wherever timber is cheap, are adopted here. All the shafts are timbered with regular cribs, resting one upon the other. The logs used vary in size, but average 10 to 12 inches in diameter. They are usually sawed on the face only, the other three sides being left round, but the bark is always removed. The timbers of the new McHarg shaft of the Morning Star are better made than any others on the property. They are shown in Fig. 4, Plate V. They are sawed on all four sides, and are 9 by 10 inches through. The 10-inch face stands vertically. The tenons on the ends are 9 inches long, and the shoulders 2 inches each, so that each set timbers one foot of shaft. The shaft is to be 9 feet by 4 feet 6 inches in the clear, so that the longer timbers have the former distance between the tenons, and the shorter the latter.

The timbers are put in in the following manner: when a convenient depth has been reached below the lowest timbers—8 or 10 feet on an average, but varying according to the nature of the ground—notches are cut into the sides at each end, and cross-pieces, prolonged beyond the tenon, as in *b*, Fig. 4, Plate V, are fitted into them horizontally. Then the sets are built upon these until those above are reached. Every other set is wedged in place firmly by blocks or wedges driven back of it, and any large spaces behind are filled with blocks. Fig. 1, Plate VI, represents a section of shaft showing the way the long cross-piece is put in. The shaft is divided into two compartments by heavy planking put across and held in place by 3- or 4-inch scantling, which is spiked to the timbers. Four-inch planking at least should be used for this purpose, as anything thinner is apt to be bent in and broken, needs constant repairing, and may cause accidents by catching the bucket as it ascends. The new shaft is to have 6-inch division planking.

When a shaft reaches the contact a drift is always run from that point. If there is but this one level the timbers are supported on four heavy posts, although the method of timbering and the pressure of the ground against the cribs prevent very much weight from settling upon them. Each pair of these posts rises from a sill and is surmounted by a cap. They are similar to the square sets to be described later on. If there is a sump it is timbered like the shaft, these cribs starting below the sets at the bottom of the drift. When the level is above the bottom the arrangement may be the same, except that the posts then set a little further out so that they may rest upon firm ground. Sometimes the cribbing does not stop, but runs all the way down. A place is then cut in the side and a frame put in for the landing at the first level. The height of these sets varies; where there is no incline they are generally 6 feet high.

CLOSING THE LEVELS.—When an upper level is not in use it is closed by a door. The door most commonly used, and by far the simplest and most convenient, is arranged as follows: two strong hinges hold it to the edge of the sill of the set at the shaft, and, as the length of the door is made greater than the height of the set, the top rests against the side of the cap towards the shaft when the level is closed, and is held there by a counterpoise. When the level is in use the door is thrown back against the opposite side of the compartment, and its weight is then sufficient to keep it there. The bottom formed by such a door has a slope of about 60°, and the bucket, on striking it, slides into the drift. The face of the door has longitudinal strips of boiler-plate riveted on it to keep the bucket from tearing the wood. Where an incline is used and a truck has to be run under the shaft, as at the first level of the Morning Star main shaft, the door is made shorter, so that it just reaches across the shaft and rests horizontally on a support fastened to the other side. The truck then runs up on a track which is spiked on the back of this door.

THE DRIFTS.

Drifting commences from the contact. The drifts are either horizontal, *i.e.* drifts which run along the strike, or inclines which follow the dip of the porphyry. A level drift does not change its height on running into a shaft. The set in the shaft is the same height as the rest. When there is an incline with a track the first few sets from the shaft are higher, and gradually run down to the normal height of the drift, so that a more gradual turn for the rope may be obtained. Owing to the loose nature of the ground and its great tendency to cave, the drift timbers have to be very substantial, and, although they are obtained at reasonable prices, they form, after the labor and smelting accounts, the largest item of expense. The lumber and timber used is all pine, the only tree that grows in great abundance about Leadville.

COST OF TIMBER.—The timber is bought as round logs, and contracts are made for its delivery at the mines. The following were the prices paid by the Evening Star mine for logs delivered during October and November, 1882:

Logs 10 feet long and 10 inches in diameter at small end.............$0 60
"   12   "   " 14   "   " 1 35
"   14   "   " 14   "   " 1 50
"   14   "   " 16   "   " 1 80

The Morning Star was having delivered about the same time the following sizes:

Logs 12 feet long and 10 inches in diameter at small end.............$0 75
"   14   "   " 10   "   " 0 80
"   16   "   " 10   "   " 0 90

The annexed table of prices from a contract of the latter mine made in November, 1881, is more complete:

Logs 14 feet long and 12 inches in diameter at small end.............$1 40
"   14   "   " 14   "   " 1 75
"   14   "   " 16   "   " 1 85
"   12   "   " 12   "   " 1 20
"   12   "   " 14   "   " 1 50
"   12   "   " 16   "   " 1 60
"   16   "   " 16   "   " 2 10
"   12   "   "   8   "   " 0 30
"   14   "   "   6   "   " 0 30
"   14   "   " 10   "   " 0 90
"   14   "   "   8   "   " 0 50
Lagging 16 feet long and 4 to 6 inches in diameter............. 0 25

All the logs brought to the mine for delivery must be sound and good according to contract. They must also be full length. They are carefully inspected before unloading, and any logs crooked, unsound, short, or in any way failing to comply with the terms of contract are rejected. A certain time is set during which the logs are to be delivered, and generally a given number has to be brought each week. Weekly payments of 75 per cent are made for logs delivered and the balance is paid when the contract is fulfilled. All the mining-timbers are made from these logs.

TIMBERING DRIFTS.

In timbering drifts the ordinary set is used, consisting of a sill, posts, cap and braces. Besides these, wedges are used to hold them in place, and lagging to hold up the loose dirt. A brief description of these pieces may be advisable, as they have to be carefully made.

THE POSTS.—These are made of various sizes. Heavy or light ones are used according to the nature of the ground and the amount of use the drift will be put to, but the same sized timbers are used throughout. Generally the logs are sawed on the face only, and are left round on the other three sides. A tenon of 2 inches is made at the top, with shoulders on the front and the two sides. The front shoulder is always 2 inches deep; the side shoulders only about 1, but vary according to the size of the log, as the tenon has the fixed width and the space left over on either side forms the lateral shoulders. Thus if a 12-inch post were to be made from a 14½-inch log, the tenon would be just 12 inches across and 1¼ inches would be left for each side shoulder. The faces of the tenon are carefully squared, and the top made square with the front face of the post. The bottom of the post has no tenon, but is sawed off square. The length varies according to the use to which the drift is to be put. For an ordinary track-drift the posts are 6 feet 4½ inches long over all. A post is shown, drawn to scale, in b, Fig. 1, Plate V. It is made from squared timber, as is sometimes done, but does not differ from the round posts in any other way.

THE SILLS.—These are not, as a rule, as heavy as the posts; they are often sawed square, and always on the upper and lower sides. They are usually 8 inches thick and 12 inches wide when made for a 12-inch post. They have a 9-inch tenon at each end, with one shoulder of 2½ inches on the upper side. The posts rest upon these tenons, and the faces come close against the shoulders. The length between the tenons is generally 4 feet. Fig. 2, Plate V, shows a sill.

THE CAPS.—The caps are made of logs corresponding to the posts in size. When the latter are not square, neither are the caps, but are then sawed only on the lower side. They have a tenon at each end which is 9 inches long on the under side. On this side the collar is 2 inches deep. When this tenon rests on the top of the post, the shoulder of the cap comes against the tenon of the post and the face of the cap rests against the front shoulder of the post. There are shoulders on the sides of the cap also, but these are, like those on the post, seldom over 1 inch deep. They start 2 inches back of the lower shoulder; that is, 11 inches from the end of the cap. The length of the cap, like that of the sill, varies greatly. Usually it is 5 feet 10 inches over all; that is, 4 feet in the clear. *a*, Fig. 1, Plate V, represents a cap. It is turned lower side up to give a better view of the tenon.

COLLAR- AND FOOT-BRACES.—These are made of sawed timber. They are 8 by 10 inches thick, and of varying length. They are sent down the mine in long pieces and cut of the length required when the set is put together. The collar-brace rests on the lateral shoulders of the posts and against those of the caps. The foot-braces lie on the floor of the drift between the posts.

Fig. 3, Plate V, shows three sets of drift timbers of about the usual strength for an ordinary track-drift. From this figure it may be seen how the various timbers fit upon one another. It must be remembered that in the ground the space between the sills is filled in, and that the foot-braces, which do not lie on any shoulders, rest upon this filling.

As accessories to the set Wedges and Lagging may be mentioned.

WEDGES.—Wedges hold the timbers in place until pressure sets upon them. They are made from any suitable material. Waste lumber or logs are sawed into blocks 16 to 18 inches long and these ripped into pieces about 4 inches square. Such a piece is then sawed from one end edge diagonally across to the other, thus forming two wedges.

LAGGING.—This is only used where the ground is loose enough to cave. It is generally required over the caps and often behind the posts. It may be of any light material, but is most commonly round sticks 3 to 5 inches in diameter. These are of any convenient length, frequently about that of two sets. The ends of the pieces are cut obliquely on opposite faces so that they form parallel surfaces. They then fit end to end and make a tight joint. Any other light material, like slabs, may be used for lagging. Everything used in drift-timbering has now been mentioned.

Certain precautions should be used in making these timbers and putting them together. The following are the most important: (1) In making the various timbers, care should be taken that they fit perfectly when put together. The tenons and shoulders should be made accurately to measurements and carefully squared. The centre-lines should be plainly marked on the inside faces so that they may be used in putting up the set. (2) In ordinary drifts the sets should stand perfectly vertical. The sill should be placed with the utmost care, as the strength and durability of the set depend greatly upon this. A trench is dug for it, and the sill then set in with the upper face perfectly horizontal, and, in a level drift, on a level with the preceding sill. If the drift has a slight inclination, say of 2 inches per set, the sill is given this difference of level. In preparing this trench, especially where the ground is not very hard, the center should be dug lower than the ends, so that the middle of the sill cannot touch. The pressure of the roof, exerted through the posts, presses the ends of the sill into the ground, and, when the ends sink, if the middle meets with resistance it will spring up and the sill will be broken, just as a stick is broken across the knee. Neglect of this precaution constantly results in the breaking of sills, this being one of the most common ways in which they are destroyed. The necessity for removing the ground that would bear at the center applies also in placing the posts and caps. Fig. 2, Plate VI, shows typically how the ground should bear on a set in a drift. (3) The lagging should be neither too strong nor too close together. Soft ground will often swell for a time after it is exposed with irresistible force. If the lagging be too close this will easily break it, while if it be further apart the soft ground will be forced into the interstices and the pressure will thus be relieved. It is for this reason, also, that round logs are often preferred to square ones in timbering winzes which pass through vein matter. Where the ground is not soft and plastic, but dry and very crumbly, close lagging may be advisable. The harm resulting from using too strong lagging is also apparent. As long as the pressure is not excessive the lagging should hold up the ground. But, since all the pressure on it is conveyed to the sets, if the lagging be very strong this may be sufficient to crush them. It is, however, better for the lagging to bend or break, as it costs much less than the sets.

COST OF SETS.—The cost of the various pieces of a set, framed and ready to be put up is given below. The posts are ordinary ones of 12-inch tenon; that is, made from a 14-inch log. Sets three feet apart from centre to centre.

	Timber.	Framing.	Sawing.	Total.
One sill	$0 67	$0 13	$0 10	$0 95
Two posts	1 50	22	10	1 82
One cap	68	40	05	1 13
Two collar-braces	30	....	....	30
Two foot-braces	30	....	....	30
Total	$3 45	$0 80	$0 25	$4 50

The distance of the sets apart varies according to the nature of the ground. Sometimes they are not more than 2½ feet from centre to centre, sometimes they are as much as 5 feet. The average is about 3 feet, which would make the cost of the framed timbers $1.50 per running foot of drift. This estimate is exclusive of wedges, lagging, and the cost of putting the sets in place, which would bring the cost very nearly up to $2 per foot.

THE STOPES.

The pressure on the sets is often enormous. It may arise from two causes, swelling of the ground, owing to access of air, or an actual caving or settling of the roof. The latter is far the more powerful source of pressure, and is more common than the other, especially under the White porphyry. When a large body of ore is being stoped along a drift the whole region settles. The caps are often crushed to splinters on the posts and broken in the middle, while the sills spring up at the center and thus almost close the drift. To support the roof in the stopes the most substantial methods of timbering have to be employed. The stope supports are simple. They are stulls and head-blocks, cribbing and square sets.

STULLS AND HEAD-BLOCKS.—These are employed as temporary supports to hold the roof during stoping until more permanent ones can be made. They may be permanent only where the stope is very small and the spaces between them are at least partially filled with waste. The stull is merely a post or log, and is made 10 or 12 inches shorter than the distance between the walls where it is to be used. The head-block is a block sawed on two opposite faces, and about 1 foot thick by 16 inches broad and 2 feet long. In setting them up a smooth place is made on the floor for the stull, and on the roof for the block. The stull is then put up and the block put over it. Wedges are driven above the block until the stull is held in position. Pressure soon settles upon it, and it is then held tightly. The stull is always put in at right angles to the roof. Fig 4, Plate VI, shows a stull in position. The wedge used to hold it up in the first place is driven from the lower side.

CRIBBING.—Cribs are the most common stope supports in the Evening and Morning Star mines, and, though the most expensive, are by far the most substantial and safe supports. Where the ore-body is very large they are almost indispensable. Cribs are generally rectangular in shape. When large, each crib or set consists of five logs, two large ones, notched at the middle and each end on both sides, and three shorter ones, notched only at the ends. Two long logs are laid down, and three short ones are placed across them, the notches fitting into each other; then two more long ones, and so on up, set above set, until the roof is reached. If the platform on which they rest be not horizontal, an extra log is put under the lower side as in the figure. The size of the logs varies, but commonly a 10- to 12-inch log is employed.

The cribs are carried to the roof, and the compartments in them are carefully filled as full as possible with waste, as this helps to keep the cribs in shape and to support the pressure. It is also well, if the cribs are very high, to fill the spaces between the logs with small timber, as this will receive a part of the pressure when the cribs settle and remove that much from the ends of the logs. Some of the cribs of the Evening Star are between 70 and 80 feet high, and they are found to be the only thing that will support the roof in such places. Fig. 3, Plate VI, shows the form of a set of cribs. The foundation and roof have been smoothed off, and the cribbing is built up vertically although the floor slants. The compartments are completely filled with waste.

SQUARE SETS.—This form of stope timber is seldom used in either mine, especially in the Morning Star. The square sets used here differ from those in use on the Comstock Lode and described by J. D. Hague, and are apparently not so good a form. They differ in no essential way from the ordinary drift set already described, only the post has a shoulder on the fourth side for an additional collar-brace. One set is placed directly upon the other, the cap of the lower forming the sill of the one above. Where the ore-body is large and regular this form of support is not now used. Where it is very irregular and narrow, but runs some distance vertically, it is always used. In stoping, if square sets are to be used, a face is started as wide as a set, and extending from the top to the bottom of the ore, and this is run forward through the block like a great high drift. But if cribbing is to be used in a large body they start at the bottom, dig under the body and prepare a foundation of proper shape and size, and then work up, stoping overhand and taking out a pillar. The cribbing is built up as fast as the ore is taken down, and any waste is utilized for filling up the cribs. In such places it is often difficult to get waste for this purpose.

For making all these timbers each mine is provided with a saw-mill and framing-rooms. In both cases they are connected with a shaft-house, and the same boiler supplies the steam for the hoister and the engine which runs the saws. Only plain sawing, cutting into lengths and making wedges is done by machinery; all the rest of the framing is done by hand. As soon as the timbers are sawed they are taken to the framing-room, framed and stored in piles of pieces of the same kind and size, ready for use. From here they are distributed to the various shafts of the mine as needed. The crib is either notched on top with an adze or is sent below of the desired length and notched when used. The log-yards are just above the saw-mills, so that the logs are easily obtained. A team brings them to the mill and distributes the framed timbers.

The cost of drift timbers has already been given. Cribs cost according to the size and length of the logs from which they are made. Any pieces cut off the regular length are made into wedges or head-blocks, and the expenses of framing them are small. The timber for one crib of two 14-foot logs and three 10-foot logs, all 10 inches in diameter at the small end, will cost $3.60. Good lagging costs about $1\frac{1}{2}$ cents the running foot. Wedges cost, ready made, a little over 1 cent apiece, head-blocks about 10 cents, and stulls 6 cents the running foot.

SHAFT-HOUSES.

Although the arrangement of several shaft-houses on both mines is more compact, that of the Lower Waterloo has other and more important advantages, and, though not so large, it is in many ways better than any of the others. It is more complete and has machinery superior to that of any shaft on Carbonate Hill, and the arrangement of the dumps is also very good. It has therefore been selected for description.

MACHINERY.—The hoisting-engine was made by J. W. Jackson, of Denver. It is a powerful one and capable of much more work than it is now required to do. The spool is 4 feet in diameter. It is worked by friction gearing, as are most of the hoisters of the camp. The large friction-wheel attached to the spool is 6 feet in diameter and works against a 2-foot paper-tired wheel. The axle of the wheel rests in eccentric sockets, so that by turning the latter one way the large friction-wheel is thrown against the small, while by turning the other way it is thrown against the break-block. The spool rests on a solid wooden frame which is on a massive masonry foundation. The rope is a 1-inch steel wire cable. It passes to the sheve over a small pulley which revolves on a bar along which it can slide, keeping on a line with the sheve and the point where the rope binds on the spool. The sheve is 4 feet in diameter, and is supported on a strong framework over the hoisting-compartment, which is carefully braced on the side towards the hoister.

There is a fine air-compressor on the same floor. It was made by Sargeant & Cullingworth, of New York. It was put in for running a pump at the bottom of the incline and for using air-drills, but so far has been used exclusively for the former purpose. This air-pump conveys the water to a sump at the foot of the shaft; from here a steam-pump of peculiar and not very satisfactory pattern raises the water to the surface, a distance of about 200 feet. The steam is supplied by two boilers. At present one of these is able to do all the necessary work, so they are run alternately, being changed every few weeks. The engineer attends to the engine and to firing the boilers. No firemen are employed at any of the shafts.

The water for the boilers is obtained directly from the city mains, which run immediately under the shaft-house. A large tank holds a reserve, and is filled at will by turning on the tap. The pressure in the main is considerable, and all the lower shafts get their water in this way. The upper shafts, Nos. 1, 2, 3, 9, 10 and 11, Plate II, are too high up, so each mine has to have a small steam-pump which supplies it with water daily. The water is of excellent quality, containing little or no carbonate of lime.

HOISTING-FLOOR.—The position of this shaft-house, like all the others, is with its greatest length running up and down the hill, in order that the ore-bins may be some distance above the ground, and that a good dump may be obtained. For this reason, also, the hoisting-floor is made higher than the floor of the engine-room. In this shaft-house it is only six feet higher. The shaft terminates at this floor, and everything is hoisted to it. The shaft has two compartments; one is used for pumping, and the other for hoisting. A bucket, not a cage, is used. The hoisting-compartment rises 8 inches above the floor all around. The front and back of it have a triangular piece surmounting this, which is 10 inches higher at the middle than at the sides. Two heavy doors, hinged to the sides, rest upon these pieces. A rope from each of the doors passes over a pulley about 8 feet above it, and then both ropes pass to one side and join together; this single rope passes through a third pulley and has a counterpoise attached to it. Hence when one door is pulled open or shut the other is too, and at the same time they work very easily. The doors are always kept closed when the bucket is down. Behind the shaft the floor is covered with boiler-plate. From this one track leads to the ore-bins and waste-dump, and another to the "wash" dumps, on which anything hoisted too poor to pay, but rich enough to make ore by dressing, is thrown.

The bucket used for hoisting has a capacity of 6 to 8 cubic feet. Sometimes it is of wood, especially where there is no incline, but here and at the Morning Star main shaft it is iron. Iron buckets are made of $\frac{1}{4}$-inch boiler-plate, with the seams strongly riveted. They are cylindrical at this shaft, and their length and breadth are about equal. At the other shaft they are shaped like an oil barrel, and are about the size of one. The wooden buckets are of this shape, made of hard wood and bound with $\frac{1}{2}$-inch wrought-iron bands. When hoisted, the bucket strikes against the doors and throws them open. It is stopped about 8 feet above the floor. The dumper then closes the doors and fastens a hook, which is suspended by a rope from a point above and behind the shaft, to a ring at the bottom of the bucket. The engineer lowers, and the rope pulls the bucket back and overturns it into the car, which stands in position against the back of the shaft. It is then hoisted to its former position, the rope unhooked, the doors opened and the bucket again lowered into the shaft.

The car used is almost exactly like that described by J. D. Hague as used on the Comstock Lode, the only essential difference being in the turn-table, which at Leadville consists of three plates, the upper two of which are hinged together and turn on the third. The door at the end is the same in both cases: hinged at the top, and controlled by a lever in front. These cars are made at the mines, of 2-inch pine plank. They are lined with boiler-plate and firmly braced on the outside with bands of the same material.

ORE-BINS.—The ore-bins are at some distance from the shaft, and a covered passage leads to them. This gives the height above the surface requisite for a sorting-floor and ore-bins below it from which the wagons are loaded. At this shaft the bins are only four in number, two with grates and two without. The grates terminate over a sorting-floor which is 6 feet above the bins. Those without grates receive ore on the lower floor immediately, and only such ore as is not to be sorted is thrown into them. The gratings are 4 feet wide, and consist of thirty $\frac{1}{2} \times 1\frac{1}{2}$-inch wrought-iron bars standing on edge. They have an inclination of about 60°, and stop about 1½ feet above the sorting-floor. Below them the floor is guarded with boiler-plate to prevent wear. When the ore is thrown on these grates, all the fine, which is mostly granular carbonate of lead, passes between the bars to the lower floor, while the lumps capable of being hand-sorted fall on the sorting-floor. The sorter goes over all this, throwing the ore into the bins, and the gangue on the waste-dump. The waste-dump is directly beyond the ore-house. It receives only such waste as is very low in silver and lead. Any waste running over 10 or 12 per cent of lead is thrown on a separate dump to await dressing.

The other large shaft-houses are provided with ordinary Colorado hoisters and no other machinery. The Upper Waterloo and Morning Star main shaft-houses are much larger than the Lower Waterloo, and the ore-bins are differently and not so well arranged. These bins are arranged in a row along the hill so that they run across the lower end of the shaft-house, forming a sort of letter T. The car containing ore is run to the crest of the bins and then turned on a piece of boiler-plate to a track at right angles running along the edge of the bins. These bins are ten in number, five with gratings and five without. The car not only has to be turned twice every time the ore is brought up, but, unless the hill is very steep, the hoisting-floor has to be raised very high above the ground to afford sufficient dump. The Morning Star shaft has its floor about 20 feet above the ground.

All the shaft-houses have sorting-floors, and ore-sorters are constantly employed in taking out the gangue. Sorting not only lessens the amount of ore smelted, but as the percentage of lead is increased the price paid per ton for smelting is lessened, and the price received for the lead is greater per pound. An example will show this more plainly. Suppose 56 tons of ore mixed with waste averages before sorting 29 per cent lead and 35 ounces silver. If sent to the smelter it would sell as follows:

Received for lead, at 1½ cents per pound	$406 00
" silver, at $1.07 per ounce	2097 20
	$2503 20
Paid for smelting, at $14 per ton	784 00
Received from smelter	$1719 20

If the ore had been sorted and brought down to 50 tons, assaying 32 per cent lead and 38 ounces silver:

Received for lead, at 1¼ cents per pound	$480 00
" " silver, at $1.07 per ounce	2033 00
	$2513 00
Paid for smelting, at $13 per ton	650 00
Received from smelter	$1863 00

The practice of putting all the waste raised that is capable of making ore by dressing on separate dumps is now in use at all the shaft-houses on the Morning and Evening Star, and both mines now have large quantities of this material on hand. The third-class dump of the Upper Waterloo gave, by assay from a number of samples carefully taken, from 27 to 30 per cent of lead and from 4 to 6 ounces of silver per ton. This dump is higher in lead and lower in silver than any of the others.

UNDERGROUND WORKINGS.

DRIFTS.—All the shafts worked either strike the ore or come very near it. As soon as contact is reached development is begun by running exploring drifts along the contact. They are either horizontal or have a decided dip. Such drifts may be divided into two classes; (1) regular highways which collect all the material to be hoisted and convey it towards the shaft, and (2) numerous branches of these which run off developing the ground, each serving at most as the outlet for only a limited territory. The former are the larger, more carefully made and better timbered. They are always supplied with a tramway along which the ore and waste is conveyed in cars or in buckets set upon trucks. The latter, being comparatively little used, are less strongly timbered, and where the ground is good, as in the Lower Waterloo, sometimes not at all. They have no tramway and are not necessarily smooth like the former. The vein-matter taken out through them is conveyed to the nearest track-drift in wheelbarrows. If a shaft or level has no incline up which the ore is drawn by the hoister, only the ore on or above that level is taken out by it, as the dip is too steep to move the ore up hill by hand. If there is such an incline the ore may be taken out for an indefinite distance below. The Morning Star main shaft, No. 2, Plate III, and the Lower Waterloo shaft, No. 6, Plate III, are the only ones that have such inclines. In both cases the incline serves as a main outlet, and side-drifts run out from it horizontally. Cross-drifts are again run from these, dividing the ground up into blocks and fully exploring it. The Morning Star incline is the better of the two. It begins at the point where the shaft strikes the contact and, following the contact, runs to the end of the property. It is not shown on the section. This incline is perfectly straight and substantially timbered with 12-inch sets (14-inch logs). A single track runs down it. The first 15 feet from the shaft is horizontal, though the sets are higher, so that the top keeps the regular slope of the incline. The set in the shaft has a 10-inch drum at the top. When a bucket is to be sent from the surface down the incline it is lowered to the level. The trammer has a low truck at the foot of the shaft to receive it. A chain just of the right length is fastened to the front of the truck. The trammer receives the bucket, guides it to its proper position on the truck, sees that it rests with its ears across the drift, and hooks the chain over the rim. He then pulls the truck over the crest of the incline and jumps on. The engineer, having merely paused at the level long enough for the bucket to be properly adjusted, now lowers it to its destination. On hoisting, the trammer comes upon the truck and loosens the chain just after the crest is reached, thus leaving the bucket free to go up without stopping. The rope used is hemp, which is much better for this purpose than wire. It bends over the drum in the set at the foot of the shaft, and over rollers at any point along the incline where the slope changes. At various points on either side of this incline horizontal drifts have been run, and from these smaller drifts. All the ore is brought to the incline through these, and there transferred to the bucket. Some of these drifts on the south side are outlets of quite extensive workings, and are provided with tramways and cars.

The arrangement at the Lower Waterloo is much less convenient, and developing work only is pushed on down this incline. It is intended to take out the ore through the new McHarg shaft, which will strike this contact below the present workings. A level drift runs from the second level of the shaft out 200 feet to the south-east till contact is reached, and here a very steep incline begins, so steep that it is not safe for a man to ride down it. A special truck and bucket have been devised for use on this incline, the rope being fastened to the truck.

The Evening Star upper shaft has also an incline, but with a very moderate slope, and the ore about it was hauled to the shaft by a donkey kept underground for this purpose. All the other shafts and levels work only the ground at or above their own level, and long horizontal drifts run from these to collect the ores from the contact.

CHUTES.—In a stope the ore is thrown or wheeled to the nearest track-drift below to be taken out. Instead of throwing it on the floor of the drift, where it would be greatly in the way and from whence it would have to be shovelled into the cars, the ordinary chute is used when possible. These extend either from some accessible point in the stope to the track-drift below or, more commonly, from a drift which in turn leads to the stope. Chutes are from 3 to 6 feet wide at the top and $1\frac{1}{2}$ feet at the bottom, and are about 1 foot deep. They run from the bottom of the upper drift to the side of the lower, and protrude enough for the car or bucket to be pushed under them. A sliding door at the bottom closes them so that the contents will not fall out. If two drifts cross one above the other a chute may be run from the bottom of one to the top of the other at the point of crossing. The Upper Waterloo has an excellent double chute of this kind. It runs from a track-drift which is the outlet of a great deal of pure carbonate of lead to the main double-track drift running from the second level of the shaft. At the place where they cross a broad chute is run from each side of the upper drift into the top of the lower. As the drifts are close together the chutes are short, but being very broad they hold a great deal. At the end of each chute there are two doors, one over each track. One chute is used exclusively to receive ore, the other waste. It is often desirable to have a double chute where the drifts do not cross, so that ore and waste may be sent down and yet never go into the same receptacle. For this end a very broad chute is made with two doors at the lower end, and a middle partition which runs nearly to the top. On this partition a door is hinged which swings from one side to the other. Thus when one side is in use the other side is closed. If the chute starts from the bottom of a drift instead of the side, the door is hinged to the bottom along the middle line of the chute.

WINZES.—In the Evening Star the ore-bodies were so thick that the main drift for conveying the ore to the shaft would be 20 and even 60 to 80 feet below the contact. In such cases winzes are sunk from the upper workings to the lower, down which the ore is thrown. To avoid accidents winzes are generally made at the ends or sides of drifts. They are timbered with cribs just like shafts, only the logs are lighter. They are tightly lined with planking running up and down to prevent the rocks from destroying the timbers.

STOPING THE ORE.—When new ground has been thoroughly developed the ore, if in considerable quantity, is left standing in reserve until the management is ready to take it out. In the mean time everything is prepared so that it may be removed in the cheapest and most expeditious manner. They generally begin to stope on the lower side of one or more blocks, and work upward along the contact. The manner of stoping varies slightly according to the nature of the body. If the ore is damp and soft, like the ores of the Forsaken and Old Waterloo, it is picked down only. When the ore is badly mixed with waste, picking is also done as much as possible. Where the ore is hard, or where it is quite solid but unmixed with waste, it is blasted out. The highly siliceous ores are the hardest to stope, as they are very hard to drill and often break with a short fracture. The thick bodies of granular carbonate of lead are also blasted down. In the Upper Waterloo the holes were bored in this ore in a few minutes with long iron augers, and a shot put in. The ore was thus brought down much faster than a man could pick it. The augers were used instead of drills, as the powder made by the drill packed before it and made the progress very slow, while with the auger a 1-inch hole 2 feet in depth could be bored in a few minutes. As the ore is stoped it is separated from the waste as much as possible and sent up to the bins, where it is again sorted by daylight. As much of the waste as can be stowed is kept down the mine; the rest is sent up and thrown on the dump. Anything fit for dressing which has to be stoped is sent up for the third-class dump.

The method of putting in the stope-supports is very simple when the ore-body is not exceedingly thick; that is, when it is less than six or eight feet thick, as are most of the bodies of the Morning Star mine. In such a body stulls are put in for temporary support as the work progresses, and as soon as there is room these are replaced by cribs for permanent support. The waste is stowed in these cribs. Where the body is much thicker and stulls cannot be used, square sets or, much more commonly, high cribs are put in at once. The way these are put in has already been described.

ASSAYING THE ORES.—The variety of the ores, their differences in value, and the fact that many of the products, though alike in all other respects, are sometimes ore and sometimes waste, have already been shown in the former chapters. Unless the percentage of lead in the ore is very high, or the silver is visible in it as the chloride, the most experienced miner cannot depend on looks alone to distinguish it from waste. The only trustworthy way of distinguishing the ores is by assay. The importance of having frequent assays made to tell what is ore and what is waste cannot be overestimated. The old Morning Star Company did not pursue this course, and as a consequence passed by large quantities of ore, and, worse than this, stowed away large quantities of it because it looked like waste, when in reality it would at that time have paid them handsome profits. To keep track of the ores each mine is furnished with an assay office, an assayer and his assistant. Every "shift boss" is required to collect frequent samples from wherever development is progressing and from wherever his men are stoping, so that he may never miss any ore. If there is any doubt about the value of any material, even when the quantity would not exceed a ton, an assay is made to decide upon it. The superintendent and time-keeper also sample where they think necessary, and to the latter officer belongs the duty of recording the samples and distributing the returns of the assayer. The ore-sorters also send in doubtful samples to assist them in sorting. Consequently there is little danger of passing over an ore or of mixing it with much waste. By a few assays from one place the miner can soon learn to tell ore from gangue at that place.

Each assay office has, on an average, from 40 to 50 of these assays a day. Of course great accuracy is not required, as it is only desired to distinguish ore from waste. One assay of each sample is made. Both the lead and silver are determined. The lead assay is made first, and if the button shows over 10 per cent of lead it is cupelled for the silver; if less than this, a scorification is made. Where the per cent of lead is low there is no sulphur, and it is found that the lead collects almost all the silver when there is not much of the latter metal present. There is considerable loss only where there is a great deal of silver, and in such a case enough can always be collected to show that the ore will pay. The best furnace for these assays is the soft-coal furnace with two muffles, one above the other. Such a furnace when hot will allow seven to nine crucibles to run in the upper muffle, while twelve scorifiers run in the lower. The heat can be easily regulated, and the same furnace is excellently adapted for control assays.

SELLING THE ORES.

The ores are not treated by the companies, but are sold to the various smelters about Leadville and to Grant's smelting works at Denver. The Leadville smelting works are situated along California and Big Evans Gulches, where water for the engines and furnaces and an approach for the railway, as well as fine positions for furnaces and ore-floors, are afforded. The ore is hauled from the shaft-houses to these works in wagons. This is done by a contractor who agrees to haul all the ore taken out at a certain rate per ton (from $0.70 to $1.00). The two mines have a weigh-house, and all the ore shipped is weighed there as well as at the smelters.

The ore is sold in lots of about 50 tons each. They seldom vary 5 tons from this amount. Each lot is sampled at the smelting works, and the ore is sold on the assay of this sample. When reduced to a final half-pint the sample is divided into three portions and sealed up in bottles or paper sacks. One of these goes to the mine assayer and one to the smelter. The third is kept for a referee in case of dispute. Each assayer makes three scorification assays for silver and duplicate assays for lead. On all ordinary lots running from 10 to 40 ounces the assayers should agree within a half-ounce. Their lead assays should come within less than $\frac{1}{2}$ per cent of each other. When the assays are satisfactory the mean between the two returns is taken as the basis on which the ore is sold.

SCALE OF PRICES.—It is highly desirable to the smelter to have ores rich in lead to mix with those poor in that metal, and a scale of prices based on the amount of lead in the ore is the result. Thus, if the ore contains very little lead a very low price is usually paid per pound for it, and a high price is charged for smelting. If, on the other hand, the ore contains much lead a higher price per pound is paid for it, and a low charge is made for smelting it. The price paid for silver is always six cents an ounce less than the market value, whatever the richness of the ore. The following is a copy of an ore contract. It shows the form of the contract and the scale of prices. The italics occupy blank spaces in the printed form and are filled in when the contract is made.

ORE CONTRACT.

THIS AGREEMENT, made this *fifteenth* day of *July*, 1882, between *The Morning Star Consolidated* Mining Company, party of the first part, by W. S. WARD, General Manager, and.............................. *Smelting Co.*, party of the second part, WITNESSETH :

1st. That said *Morning Star Consolidated* Mining Company, by its General Manager aforesaid, in consideration of payments hereafter mentioned, to be made by said second party to said first party, has agreed and does hereby promise and agree to and with said second party to sell and deliver at the works of said second party in Leadville, *three fourths* (¾) *of the output of said Morning Star Consolidated mine*, tons of ore from its mines, the delivery thereof to begin on the *fifteenth* day of *August*, 1882, and to continue (unavoidable delays and accidents excepted) until the *first day of February*, 1883. And the amounts delivered each day (Sundays excepted) to approximate *three fourths* of the product of said mine for each day, it being the intent hereof to sell and deliver to said second party only *three fourths* of product of said mine during the time which will be required to produce and deliver the said ore hereby agreed to be sold to said second party.

2d. In consideration of the foregoing, the said............................*Smelting Co.*, party of the second part, agrees to pay for the said ore, to the said first party, at the following rates :

For Silver, New York quotation at time of settlement, less *six* (6) cents per ounce.

FOR ORE CONTAINING:

Up to and including 20 per cent Lead.........................25 cents per unit,* less $16 per ton
Over 20 "	" "	25 "	" ....25 "	" "	15 "
" 25 "	" "	30 "	" ....30 "	" "	14 "
" 30 "	" "	35 "	" ....30 "	" "	13 "
" 35 "	" ...	40 "	" ....35 "	" "	12 "
" 40 "	" "	45 "	" ....40 "	" "	11 "
" 45 "	" "	50 "	" ....40 "	" "	11 "
" 50 "	" "	55 "	" ....42 "	" "	10 "
" 55 "	" "	60 "	" ....45 "	" "	10 "
" 60 "	" "	65 "	" ....50 "	" "	10 "
" 65 "	" "	.. "	" ....50 "	" "	9 "

All Gold over one half ounce per ton to be paid for at *the rate of twenty dollars* ($20.) *per ounce for entire gold contents*.

All assays, whether Gold, Silver or Lead, to be reported to the tenth of the unit.

All expenses of hauling in excess *seventy cents* per ton to be paid by said party of the second part, and all payments to be made in lawful money, upon request of said first party, whenever fifty tons or less shall have been delivered, sampled and assayed.

All ores shall be sampled and assayed without unnecessary delay, and in case of disagreement as to values, as shown by the assays of the respective parties hereto, the umpire or third sample, shall be referred to................................for assay, and his determination of the value from such sample shall be final, and fix the value of said ores so sampled, which values so fixed shall thereupon be paid for said ores by said second party to said first party, and the expense incurred for said umpire assay to be paid for by losing party. Moisture determinations to be made *in case of dispute on the water-bath*.

It is understood and agreed that if for any reasonable cause the said first party should be unable to produce the amount of ore herein contracted to be sold, then, as to such part not produced, this contract shall be void.

Signed { ....
         { ....

* The unit is one per cent of a ton; that is, 20 pounds.

Since February 1, 1883, the following very advantageous terms have been made by both mines with one of the smelting companies: For silver, 6 cents an ounce less than the New York quotation at the time of settlement. For lead, 45 cents per unit when the New York price is above 5 cents per pound, 40 per unit when it is between 4½ and 5 cents, and 35 cents per unit when it is below 4½ cents per pound. From this value thus obtained the smelter deducts $9 a ton as a charge for treatment.

VALUE OF THE ORES.—The ore-books, recording the ore-sales, do not state the amounts of silver and lead in the ores, though they give all the data for obtaining them. To obtain these, tedious calculations have to be made for each lot. To give a general idea of the average richness of the ores of the two mines the following table was prepared. It shows the number of net tons of ore shipped from each mine, and the silver and lead in them, for each of the three months, August, September and October, 1882:

TABLE SHOWING PRODUCTION OF EVENING STAR.

	Net Tons.	Oz. Ag.	Tons Lead.	Av. Oz. per ton.	Per cent Pb. per ton.
August	2,828,089	115,725.3	574.589	40.92	20.31
September	2,228,515	88,278.2	302.636	39.65	13.58
October	1,258,871	46,117.4	140.215	36.63	11.14
Total	6,315,475	250,120.9	1,017.440		

TABLE SHOWING PRODUCTION OF MORNING STAR.

	Net Tons.	Oz. Ag.	Tons Lead.	Av. Oz. per ton.	Per cent Pb. per ton.
August	1,749,029	43,014.0	678.848	24.53	38.8
September	1,070,921	23,753.4	473.057	22.27	44.2
October	1,975,006	52,238.0	691.968	26.45	35.0
Total	4,794,956	119,005.4	1,843,873		

These tables show prominently the high average of the Morning Star ores in lead. The Evening Star ores never averaged as high as the former in this metal, but are lower in lead for the three months given than they were before the large body was stoped. The fact that the Evening Star ores are much richer in silver than those of the Morning Star is also prominently brought out. The variations in value of the various lots is, comparatively speaking, small. Thus in the Morning Star sales for the three months only 8 lots out of the 98 went over 40 ounces per ton, and the highest was 46 ounces. The lowest ran 8½ ounces, with 4 lots below 10 ounces.

In lead the variation was great, the limits being 65 per cent on the one hand

and 14½ per cent on the other. There were 16 lots below 25 per cent, 18 lots above 50, and 6 above 60 per cent.

The Evening Star shows no greater variations. From the 125 lots sold, 8 went over 60 ounces, with a maximum of 75 ounces. But 3 lots ran below 25 ounces, and the lowest was 23½ ounces. The lead was always very low; only 13 lots assayed over 25 per cent, and the highest was but 31½. Of the 17 lots which ran below 10 per cent lead, 13 were sold during the month of October. The exceedingly low minimum was 6 per cent, the lowest of any shipment of either mine noticed on the books.

In order to show the expenses of mining and the items of greatest expense, the following table, showing the expenditures of the Evening Star mine up to Oct. 1, 1882, is given. The total represents the amount received from the smelter for the ores. The value of the silver and lead produced by the mine would, of course, be far greater.

TABLE.

Cash on hand	$4,824 09
Labor	356,059 01
Improvement	20,844 85
Machinery	17,268 53
Iron, steel, etc.	35,603 81
General supplies and expenses	4,049 18
General transportation	272 36
Assaying materials	3,324 48
Horse feed	3,957 09
Legal service and surveys	3,293 99
Office expenses	1,568 18
Ore transportation	42,333 11
Timber	41,998 29
Fuel	12,434 18
Insurance	479 15
Telegraph and telephone	723 97
Leadville Water Co.	1,834 25
Tax	477 62
New York office as net earnings	1,410,592 20
Total received from sale of ores	$1,989,998 84

THE END.

SECTION 1.

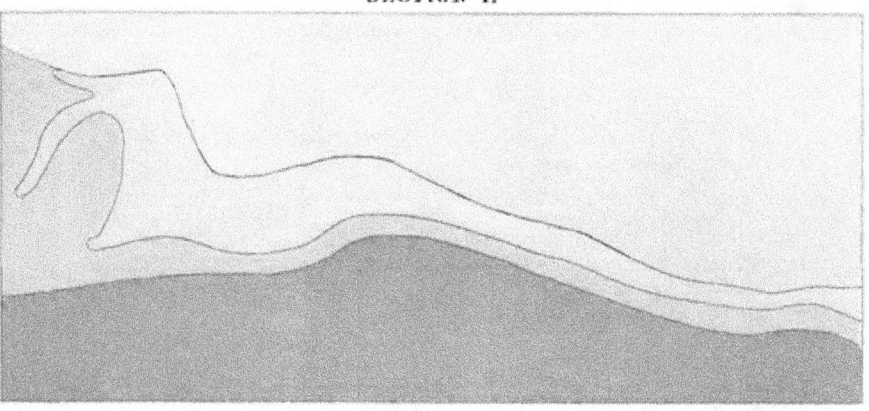

SECTION 2.

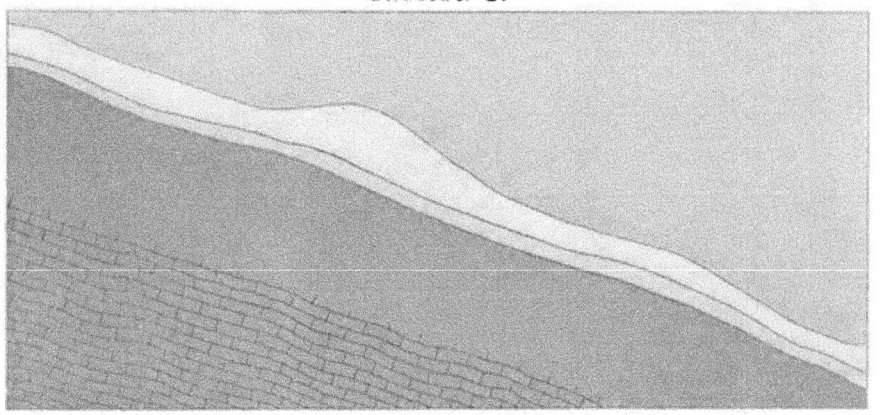

SECTION 3.

Pl. II.

PLAN OF MORNING AND EVENING STAR MINES.

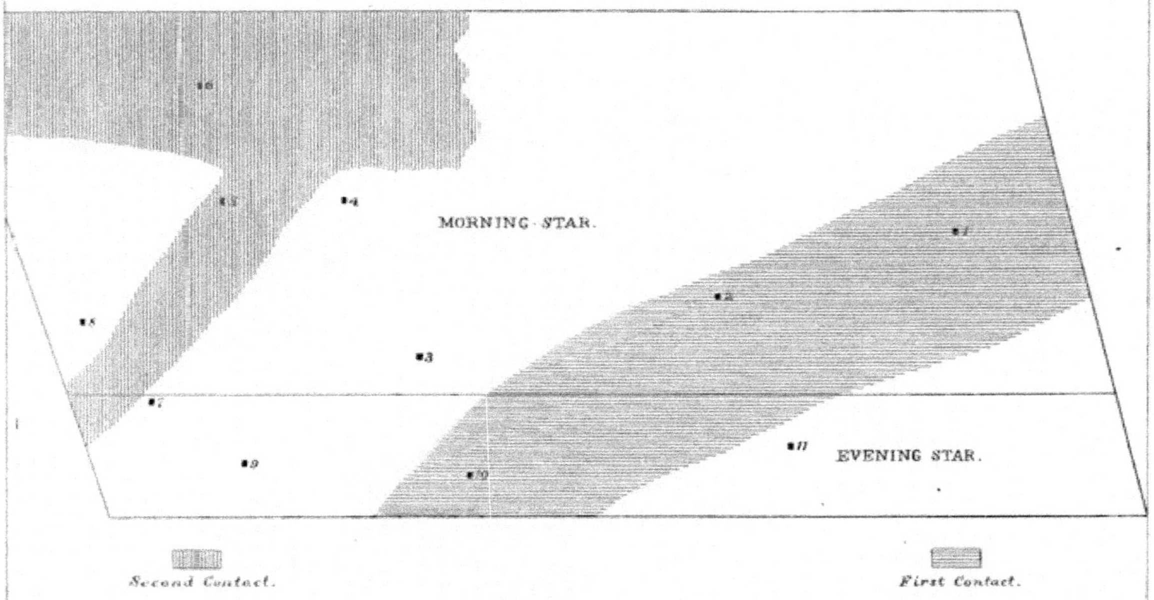

MORNING STAR.

EVENING STAR.

First Contact.

Second Contact.

Scale: 160ft = 1 inch.

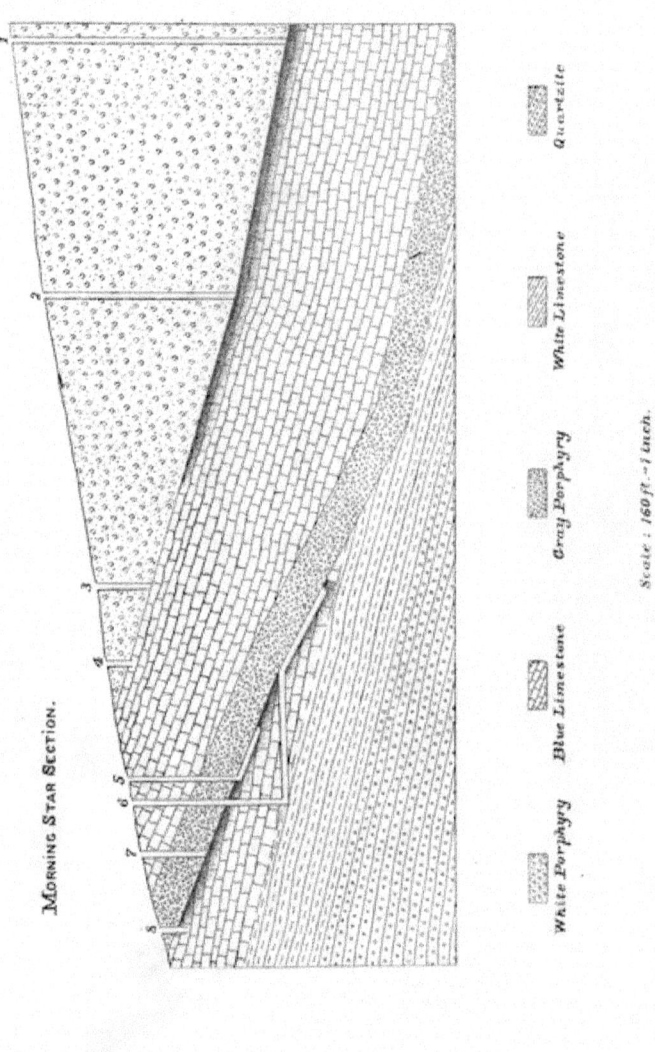

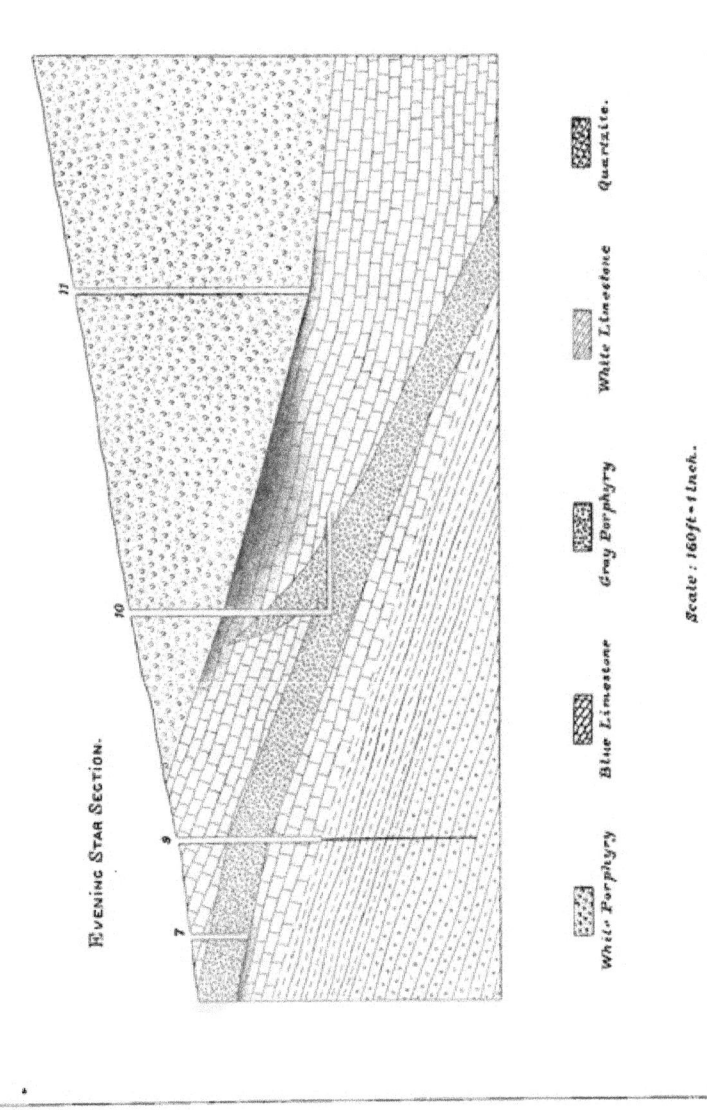

Pl. V.

Fig. 1. Fig. 4.

Fig. 2. Fig. 3.

Scale: 1:36.

Pl. VI.

Fig. 1.

Fig. 2.

Fig. 3.

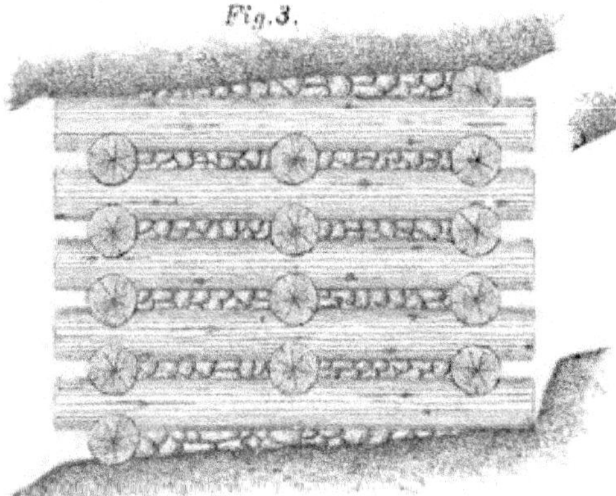

Fig. 4.

Scale: 1÷36.

www.ingramcontent.com/pod-product-compliance
Lightning Source LLC
Chambersburg PA
CBHW020237090426
42735CB00010B/1738